Electronic Troubleshooting Procedures and Servicing Techniques

Electronic Troubleshooting Procedures and Servicing Techniques

J. A. SAM WILSON

 PRENTICE HALL, Englewood Cliffs, New Jersey 07632

Library of Congress Cataloging-in-Publication Data

Wilson, J. A. Sam.
   Electronic troubleshooting procedures and servicing techniques /
J.A. Sam Wilson.
       p.   cm.
   ISBN 0-13-248428-5
   1. Electronic apparatus and appliances--Maintenance and repair.
I. Title.
TK7870.2.W55   1990
621.381--dc19   89-16351
                                                           CIP

Editorial/production supervision and
  interior design: Nancy Savio-Marcello
Cover design: Wanda Lubelska Design
Manufacturing buyer: Gina Brennan

 © 1990 by Prentice-Hall, Inc.
A Division of Simon & Schuster
Englewood Cliffs, New Jersey 07632

Printed in the United States of America

10 9 8 7 6 5 4 3 2 1

ISBN   0-13-248428-5

Prentice-Hall International (UK) Limited, *London*
Prentice-Hall of Australia Pty. Limited, *Sydney*
Prentice-Hall Canada Inc., *Toronto*
Prentice-Hall Hispanoamerica, S.A., *Mexico*
Prentice-Hall of India Private Limited, *New Delhi*
Prentice-Hall of Japan, Inc., *Tokyo*
Simon & Schuster Asia Pte. Ltd., *Singapore*
Editora Prentice-Hall do Brasil, Ltda., *Rio de Janeiro*

*To Mom and Dad, who claimed they were
not responsible for* **any** *of the mistakes
I have made since my twelfth birthday.*

Contents

PREFACE   xi

1

THE BASIS FOR TESTING AND TROUBLESHOOTING   1

Chapter Overview   1
Where Do You Start?   2
Troubleshooting with Test Equipment   3
Summary   11
Programmed Section   12
Test Your Knowledge   17
Answers to Test Your Knowledge   18

2

KNOWLEDGE: AN IMPORTANT TROUBLESHOOTING AID   19

Chapter Overview   19
DC Voltages on Amplifying Devices   20
Coupled Circuits   35
The Receiver as an Electronic System   41
Finding Trouble in a Dead Receiver   45
Operational Amplifiers   46

The Ohmmeter Test for Transistors   47
Summary   50
Programmed Section   51
Test Your Knowledge   58
Answers to Test Your Knowledge   59

3

TROUBLESHOOTING WITH A METER   60

Chapter Overview   60
About the Meter   61
Battery Terminal Voltage   70
Summary   78
Programmed Section   80
Test Your Knowledge   88
Answers to Test Your Knowledge   89

4

SERVICING WITH AN OSCILLOSCOPE   90

Chapter Overview   90
Comparison of Oscilloscope Types   91
Frequency vs. Time Domain   97
Scope Accessories   97
Measuring or Evaluating Components with an Oscilloscope   102
Amplifier Servicing with an Oscilloscope   103
Summary   109
Programmed Section   110
Test Your Knowledge   115
Answers to Test Your Knowledge   116

5

SIGNAL INJECTION AND SIGNAL TRACING   117

Chapter Overview   117
System Troubleshooting   118
Summary   129
Programmed Section   130
Test Your Knowledge   134
Answers to Test Your Knowledge   135

6

SYMPTOM ANALYSIS, DIAGNOSTICS, AND STATISTICAL METHODS   136

Chapter Overview   136
Symptoms   137
The Diagnostic   138
Summary   142
Programmed Section   144
Test Your Knowledge   150
Answers to Test Your Knowledge   151

7

SERVICING CLOSED-LOOP CIRCUITS   152

Chapter Overview   152
Closed-Loop Circuits for Voltage Control   153
Feedback Circuits That Control or Establish Frequency   163
Review of Some Basic Circuits Used in Closed Loops   168
Summary   169
Programmed Section   170
Test Your Knowledge   174
Answers to Test Your Knowledge   175

8

HUNTING FOR THE CAUSES OF NOISE AND INTERMITTENTS   176

Chapter Overview   176
Types of Noises and Their Causes   177
Intermittents   186
Summary   187
Programmed Section   188
Test Your Knowledge   192
Answers to Test Your Knowledge   193

9

SERVICING DIGITAL LOGIC AND MICROPROCESSOR EQUIPMENT   194

Chapter Overview   194
Logic Systems   195
Knowledge as a Valuable Troubleshooting Aid   205

Types of Test Equipment Useful in Troubleshooting Logic
and Microprocessor Circuits 205
Microprocessors 208
An Example of a Microprocessor System 209
Summary 212
Programmed Section 214
Test Your Knowledge 218
Answers to Test Your Knowledge 219

10

REPAIRING AND REPLACING 220

Chapter Overview 220
Soldering Techniques 221
Summary 229
Programmed Section 230
Test Your Knowledge 232
Answers to Test Your Knowledge 233

11

SOME LOW-COST HOMEMADE TESTING DEVICES 234

Chapter Overview 234
Nonpolarized Electrolytic Capacitors 235
Voltages to Check Calibration 236
Using a Volt-Ohm-Milliammeter for a High-Voltage
Measurement 237
A Very Simple Logic Probe 238
A Simple Noise Generator for Use in Signal Tracing 239
Measuring AC Current with a VOM 240
A Resistor Substitution Device 241
Detector Probe 242
A Simple In Situ (On-Site) Transistor Tester 244

INDEX 247

Preface

I once wrote a test procedure for fixing an industrial electronic system. I started by warning the technicians not to operate the equipment with the safety interlock defeated. (To do so would subject the technician's body to a very high dose of radiation.) I was strongly criticized by an editor for starting on a negative note.

To this day I don't know how I would write the procedure without first warning the technicians about defeating that interlock. Maybe I was supposed to say, "Go ahead, defeat the interlock and see what you get."

Fortunately, there are no life-threatening procedures in this book. However, in addition to explaining the purposes of the book I also want to say what the book is not intended to do.

My writing experience aside, I'll start with something the book does not do. This is *not* a book on how to repair a VCR, or a television receiver, or a radio, or any other specific piece of equipment. Rather, it *is* a compendium of troubleshooting tests, measurement procedures, and servicing techniques. There is also some information on how to make repairs and replace components.

For the purposes of this book, the hierarchy of electronics—from the simplest to the most complicated—is given here:

Component.   Separate parts that combine to make an electronic circuit: resistors, capacitors, and inductors.

Circuit.   A combination of components that is used to perform a specific function: oscillators and power amplifiers.

System.   A combination of circuits: radios and computers.

In the book I discuss tests, measurements, and troubleshooting for all of these categories. For *components* and *circuits* there is a wide variety of examples. However, depending on level of schooling and experience, there are very few *systems* that all readers would be acquainted with.

If I used a television receiver, or a VCR, or an industrial control system as an example, there would be readers who are not comfortable with the subject. So I chose a radio for discussing systems. For those who would like to review the basic radio system, I have included a short discussion of a typical radio in Chap. 1. The purpose of every component in that radio is included. If you know what a component is *supposed* to do, you will be better able to recognize the problem when it is not doing its job. (That is an opinion—not a hard and fast rule.)

Because of the wide variety of problems that can occur in electronic equipment, I seriously doubt if it is possible to have a universal, fits-all troubleshooting procedure. (That is another opinion.) I know that many authors have tried to formalize a standard approach, and most of those approaches are well thought out. I have never been able to make such an approach work for myself.

However, I know that some of my readers would be disappointed if I didn't at least try, so here is my recommended procedure:

- I think the most important step in troubleshooting is to learn as much as you can about electronics. I presume you know that subject, but a little review is always a good idea. So I review some of the basics as they apply to troubleshooting in Chap. 2.
- As discussed in Chap. 1, start by looking for obvious possibilities.
- If there are no obvious faults, measure the power supply voltage. If it is not the correct value, don't go any further until you find out why.
- Use symptoms whenever possible to zero in on the section at fault. This method is discussed in Chap. 6. If symptoms don't get you to the trouble region, use a signal tracing or signal injection method, as discussed in Chap. 5.
- If these methods fail, use diagnostics, as discussed in Chap. 6.
- Once you have zeroed in on the troubled circuits, use the methods discussed in Chap. 3 and 4 to locate the defective component.
- Tough problems (like closed loops, distortion, noise, and intermittents) often require special techniques like those discussed in Chap. 7 and 8.
- Make efficient use of your test equipment, as discussed throughout the book, especially in Chap. 3 and 4. Specialized equipment that you build yourself can be very useful for locating defective components. Chapter 11 describes a few useful testers you can build.
- Digital and microprocessor circuits require special equipment and a few special techniques, as discussed in Chap. 9. However, some of the procedures are basically the same as those discussed in the earlier chapters.
- When looking for bad components, it is a good idea to keep in mind that troubleshooting often involves these steps:
  Make a measurement.

Compare the measured value with what you are supposed to get.

If the measured values do not match what you are supposed to get, find out why. The book describes specific measurements used for locating faulty components, circuits, and systems.

- When you have located the defective component, do a professional job in replacing it. Some new techniques are needed for surface mounts. Chapter 10 discusses replacement of components.

I realize that this procedure has a lot more than two or three steps, but it has as few as I can make it.

I believe your best troubleshooting help will always be your knowledge of how things electronic work. I have assumed you can use a voltmeter, an oscilloscope, a signal generator, and a frequency counter for their basic purposes. For example, it is assumed that you can use a voltmeter (or VOM) to measure a voltage. Use of test instruments for basic purposes is not a subject for this book.

The book is organized by techniques rather than by an organized troubleshooting procedure. In some cases those techniques overlap chapters—for example, should the signal injection technique be combined with the signal generator applications or with the system analysis discussions.

After spending hours debating questions like that, I decided the most important thing is to get the techniques into the book. (I'm sure the readers can put them in their own favorite categories.) However, you will see a general drift of ideas from component-to-circuit-to-system analysis whenever possible.

The selection of test equipment depends on the amount of service and the type of service done. As with universal troubleshooting rules—which probably don't exist—it is likely that there is no universal list of test equipment. So as not to be a disappointment to the reader, I will make a valiant effort to make a list anyway:

- Digital volt-ohm-milliammeter
- Analog volt-ohm-milliammeter
- Logic probe
- Dual-trace, triggered sweep oscilloscope (50 mHz or higher preferred)
- Function generator with a VCO sweep provision
- Low distortion sine wave generator (audio range)
- Transistor tester
- DC power supply (stiffly regulated) +5V, +12V

Each chapter in this book starts with an overview of the subject and a list of specific objectives. Following that, the subject matter is discussed in detail.

There is a *programmed section* in each chapter. The programmed section is *not* just a review of what has been covered; some new subject matter is also included.

A self-test, called Test Your Knowledge, at the end of each chapter gives the reader a chance to demonstrate his or her understanding of the subject matter. For those who use the book for self-study, the answers are given to all questions.

I gathered many of the ideas in this book from technicians who attended my lectures. Some technicians have written details of their favorite techniques in their letters to me. In some cases, they are ideas that I put to use in my own troubleshooting experience. If I included anyone's name I would have to include everyone's name. Frankly, I can't remember all of those names. Those of you who have helped—you know who you are—have my deepest gratitude.

I am also indebted to company training brochures and other training programs.

I have learned much from the students who have passed through my theory and lab classes. Many were working in electronics and attending school at the same time.

I would like to thank Greg Burnell and the many other Prentice Hall people who have helped to get this book into production.

Last, but certainly not least, much appreciation must be given to my wife Norma who helped all the way. She is more than a typist. In my company I am the president and she is everybody else.

J. A. Sam Wilson

Electronic Troubleshooting Procedures and Servicing Techniques

1

The Basis for Testing
and Troubleshooting

CHAPTER OVERVIEW

There is no such thing as a perfect troubleshooting procedure that works for every occasion and for everyone. If there were such a thing, every good technician in the country would be using it. Instead, each technician has a favorite approach and favorite test procedures that have been developed over years of making a living in the servicing business. However, certain basic procedures are similar in all testing, and they are the subject of this chapter.

Objectives

Some of the topics discussed in this chapter are listed here:

- How much can you rely on symptoms as a troubleshooting aid?
- What are some key words to listen for when an owner or user is describing symptoms of faulty equipment?
- How can a milliammeter be used to check a power amplifier?
- What are the statistical chances of a component being bad?
- How is a resistor sometimes used as a fuse?
- What are some of the factors that must be considered when analyzing a qualitative measurement?

WHERE DO YOU START?

A good way to start troubleshooting is to give everything a close visual inspection. This can be done very quickly. It is a good idea to look for such obvious problems as burned spots and places where a high voltage arc has occurred. Also, inspect the fuses or circuit breakers.

Be sure to look at the screws holding the back and chassis. Marked-up screws indicate that an unprofessional person may have been tinkering with the system. That, in turn, may indicate a major troubleshooting problem.

Symptoms—An Introduction

Symptoms may be useful for locating the general area of the problem. The symptoms are sometimes described by the equipment owner or user. In an industrial electronics plant, it may be the foreman of the division who uses the equipment. In consumer electronic equipment, the description is often from the customer who owns the equipment.

Experienced technicians have learned to be wary of users' and owners' descriptions of symptoms. Because those people are usually not electronic experts, they are likely to describe a problem in terms of things they are already familiar with. Here are some key words to listen for: fire, smoke, overheating, the smell of something burning.

The most reliable symptom analysis is given by the technician after energizing the equipment. Technicians know that certain symptoms in a system usually mean that a certain component has failed. That failure has often been observed.

Some technicians keep track of recurring problems in a personal notebook. In the case of consumer electronic systems, the symptoms—and cures—are available on computer data bases.

Symptoms are indeed a valuable guide, but it is not a good idea to base a complete troubleshooting procedure on symptoms alone. For example, distortion in a radio's output sound can be caused by a low terminal voltage of an aging battery. In fact, that is very possible. However, distortion can also be caused by a transistor that is no longer able to do its job. More will be said about symptoms in Chap. 6.

The Statistical Approach

Although it is not often discussed, experienced technicians make use of a type of statistical analysis when troubleshooting.

In the case of sound distortion just described, there are two possible faults for the symptom. From experience, the technician knows that a battery will fail more often than a transistor. So the battery would be checked and/or replaced as a first step.

Table 1-1 lists component failures in their order of probable occurrence.

Because it is usually easier to replace a battery than a transistor, ease of replace-

TABLE 1-1 COMPONENT FAILURES IN ORDER OF PROBABLE OCCURRENCE

Mechanical and electromechanical devices such as relays and switches. Also, plugs and sockets, especially if they are often used.

Components that get hot in their normal operation. Examples are power amplifiers and rectifiers.

Electrolytic capacitors. This is especially likely with very small versions and those subjected to high voltage.

Active devices like transistors and SCRs.

Passive devices like resistors and capacitors.

ment is also a factor. Who doesn't take the easy path first? So if there are two equal choices, you will likely try the easier repair first—even if it is out of order in Table 1-1. However, given equal difficulties it is best to start by replacing the component that is more likely to fail.

TROUBLESHOOTING WITH TEST EQUIPMENT

Having done the preliminaries, you are ready to take a serious look at the faulty equipment using your test equipment. You may find different opinions about this, but in this book the first test equipment measurement is the same for every troubleshooting problem:

Step 1: Measure the power supply voltage.

You can be sure that nothing is going to work in the system if the power supply voltage is not the correct value. Remember that a low power supply voltage can be the cause of a failure in some other part of the system. The example of distorted output sound of a radio caused by low battery voltage has already been discussed. A low power supply voltage can cause a failure of a circuit that is not located anywhere near the supply.

It is best to measure the power supply voltage while it is delivering current to the system. For example, in battery-operated equipment, the system should be energized when the battery voltage is being measured.

Figure 1-1 shows two examples of simple power supplies. The one in Fig. 1-1(a) is a simple half-wave rectifier; the one in Fig. 1-1(b) is a typical battery supply. The half-wave rectifier circuit of Fig. 1-1(a) will be used as the first example.

If there is no output voltage from the supply, check the obvious possibilities, such as a defective switch, blown fuse, or open diode.

Figure 1-2 shows quick ways to check a switch and a fuse. (Be sure to set the meter scale to a value that is high enough to accommodate the full circuit voltage or current.)

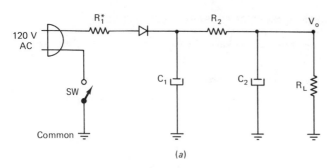

(a)

*This surge-limiting resistor protects the diode while C_1 is charging. Its resistance value is long.

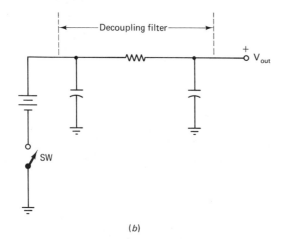

(b)

Figure 1-1 Simple power supply circuits.
(a) A half-wave rectifier.
(b) A battery with a decoupling filter.

For the connection in Fig. 1-2(a), a high voltage should be displayed by the meter when the switch is open, and the meter display should show zero volts with the switch closed. Even though the switch makes a clicking noise it may be permanently open or closed.

The connection of Fig. 1-2(b) can be used to check the fuse. If the fuse is open, you will measure the full DC current. You should know the amount of current normally drawn by the load. If not, you can set the current meter to the maximum current scale.

The meter in Fig. 1-2(c) should measure the full battery current under load *when the switch is open.* When the switch is closed, the meter reading should be zero.

In the battery supply a permanently closed switch will often result in a dead battery because the equipment has likely been on for a long time.

Suppose the power supply is delivering current to an amplifying system. The power amplifiers will require most of the power supply current. If a power amplifier is open, or cut off, the milliammeter display in Fig. 1-1(b) and 1-1(c) will show a current that is too low.

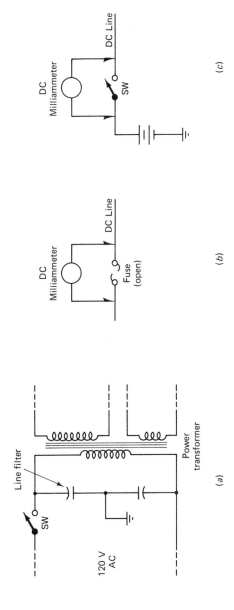

Figure 1-2 Methods of checking the on/off switch.

(a) The switch is in the primary circuit of the power transformer.

(b) A fuse is like a switch. When the fuse is open, the DC milliammeter completes the circuit.

(c) In this case, the DC milliammeter is connected across the open switch to measure the total current drawn by the circuit.

A common problem with power amplifiers is an emitter-collector short circuit. That will result in a current reading that is too high as indicated by the DC milliammeter connection.

If the current is zero for both open and closed switch positions, you can easily tell if it is due to a defective switch or some other power supply defect. The DC milliammeter will complete the circuit and you will measure an output current.

For the connection in Fig. 1(b), you can tell if a new fuse will blow if the current reading is too high. If so, find out the reason for the high current before replacing the fuse.

There doesn't appear to be a fuse in the circuit of Fig. 1-1(a). However, the surge-limiting resistor (R_1) may have a low enough wattage rating to protect the circuit. In other words, the power rating may be chosen so that the resistor acts like a fuse.

Measure across R_1 with a voltmeter—*not* a milliammeter! A high voltage indicates that the resistor is open.

If you connect a DC voltmeter across the output of the supply in Fig. 1(a), remember that the voltage (Vo) may be as high as 160V or more. The 120V AC input is an RMS value, but the capacitors charge to the peak value of that voltage.

Leaky filter capacitors can cause an excessive current through the diode. In the case of a leaky C_2, excessive current may flow through filter resistor R_2 and that will lower the supply voltage.

Excessive ripple in the output may be caused by a defective filter capacitor. Technicians often check for this possibility by connecting a good capacitor across C_1 and then across C_2. A method of checking the capacitors that is better than that bridging method is to look for excessive ripple with an oscilloscope. (Use the AC input on the scope.)

A unique way of checking electrolytic capacitors is shown in Fig. 1-3. The AC *voltmeter* is connected in series with the electrolytic capacitor. In this application the voltmeter is being used as a current meter. Excessive AC current indicates a defective capacitor. This method has the disadvantage that you have to take time to disconnect the capacitors. Also, you have to know (from experience) the allowable AC current since electrolytic capacitors are notoriously leaky; that is, they act as if they have a resistor in parallel to allow current flow.

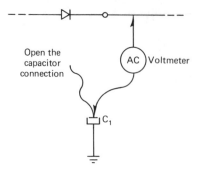

Open the
capacitor
connection

AC Voltmeter

C_1

Figure 1-3 A special method of checking electrolytic capacitors and power supplies.

If there is a low power supply output voltage, start by checking the AC power line voltage. Defective rectifiers or filter capacitors are highly probable causes of low supply voltage.

It is not rare that a low power supply voltage is being caused by a defective circuit in the system being supplied. For example, a short circuit to ground in an amplifier can cause the power supply to be excessively loaded. That in turn can cause a fuse to blow or a diode to burn out or a low output voltage.

The troubleshooting techniques described here will work for other types of unregulated supplies. More will be said about power supply troubleshooting.

Types of Measurements — Qualitative vs. Quantitative

Assuming that the power supply voltage is OK, additional measurements are in order. It is important to understand the types and limits of measurements.

There are two major kinds of measurements: qualitative and quantitative.

With a *qualitative* measurement, you must make a judgment. An example is shown in Fig. 1-4. The desired pattern on the oscilloscope screen is provided by the manufacturer of the equipment that is being serviced. You use your experience and knowledge of the system (or circuit or component) to determine whether the displayed pattern is sufficiently close to the manufacturer's specified pattern.

Qualitative measurements are very difficult to teach and difficult to learn. It would be very helpful if technicians would spend some time making qualitative measurements in equipment that is working properly. In a short while the required experience can be gained for interpreting results.

Important questions to ask when making a qualitative judgment are

- Can the symptoms of the defective system be produced by this variation in patterns?
- Is the maximum height an important factor?
- Is the width of the pattern an important factor?
- Is it possible to make any adjustments to get the waveform obtained closer to the desired pattern?
- Is it possible that a defective component is causing the variation?

Quantitative measurements provide numbers. A voltmeter measurement is quantitative. If the manufacturer's specification says that a certain voltage must not

Manufacturer's
specified scope pattern

Scope pattern
obtained during
troubleshooting

Figure 1-4 A qualitative judgment is
required to interpret the pattern.

be less than 30V, and the voltmeter displays a voltage of 30, then the measurement has given an indication that there is no trouble at that point.

In some cases it is necessary to make judgments with quantitative measurements. If the schematic diagram indicates that a voltage should be 10V and you measure 8.6V, you will have to determine the effect of the lower voltage. A voltage of 8.6V would not be acceptable for a stiffly regulated 10-V supply. However, it might be fine as the collector voltage of a voltage amplifier.

Some measurements are not exactly in either the qualitative or quantitative category. In other words, they can be both qualitative and quantitative. A good example is the square wave test. It is illustrated in Fig. 1-5. A square wave is applied to the input of an amplifier and the output square wave is observed on an oscilloscope. When used as a qualitative judgment, the technician observes the output square wave. Knowledge of possible problems, as indicated by the square wave test, is put into play.

Figure 1-6 shows some of the possibilities. A technician has to make the decision if a slightly improper square wave is still acceptable as far as the amplifier is concerned.

If you use a square wave generator with a variable frequency and adjust the frequency high enough, almost every amplifier will show some kind of high-frequency distortion. So the frequency at which the square wave is operated is a very important factor.

Some textbooks say that the square wave test can be used as an indication of bandwidth in a quantitative measurement. One suggested procedure is to raise the frequency of the input square wave until the output becomes distorted. Then the frequency of the square wave (at which the distortion first occurs) is multiplied by 7 or 9, corresponding to the seventh or ninth harmonic. This (supposedly) gives the bandwidth of the amplifier.

This is *not* a good test.

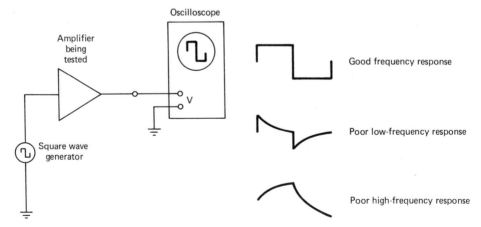

Figure 1-5 A square wave test is usually qualitative, but under certain conditions it can be used as a quantitative test.

Figure 1-6 Some examples of waveforms taken during a square wave test.

A better quantitative evaluation can be made by measuring the rise time of the square wave. Then use the following equation:

$$\text{Bandwidth} = \frac{0.35}{\text{Rise time}}$$

where the rise time is expressed in microseconds and the bandwidth is expressed in megahertz.

Figure 1-7 shows how the rise time of any pulse or square wave is measured. Note that the rise time is defined as the amount of time that it takes the wave to go from 10 to 90 percent of the maximum amplitude. You can use an $\times 10$ sweep expander (available on many scopes) to make it easier to read the rise time. (Be sure to divide the displayed rise time by 10 to get the real value.)

You have to be very careful in making this measurement. If the rise time is measured on an oscilloscope, and the scope can't pass the square wave, then the test is useless.

For example, suppose the rise time is found to be 180 microseconds, but the square wave introduced directly to the oscilloscope vertical input also shows a rise time of 180 μs. This means that all of the rise time delay in the test is caused by the oscilloscope.

In sum, you are using square wave analysis as a qualitative test if you are making a judgment on the shape of the output waveform. You are also using it as a quantitative test if you are using the rise time to determine the bandwidth.

Know Your Test Equipment

You must know the capabilities, limitations, and applications of the equipment you are using. This is best determined by carefully reading the manufacturer's equipment specifications. For example, you may be looking for glitches (undesired spike voltages). If you are using an oscilloscope that has a bandwidth of 20 megahertz, it is doubtful if you will see any glitches and thus conclude, incorrectly, that there are none. The real problem is that the bandwidth of the oscilloscope is so narrow compared to that required for displaying the glitch that you are missing. You are using a

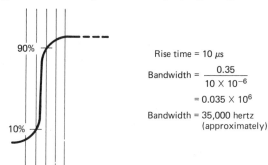

Rise time = 10 μs

Bandwidth = $\dfrac{0.35}{10 \times 10^{-6}}$

= 0.035 $\times 10^{6}$

Bandwidth = 35,000 hertz (approximately)

10% — ⊥

90% — ⊤

10 μs per division →

Figure 1-7 This illustration defines the term *rise time* as it applies to the leading edge of a step function.

piece of equipment that is not able to perform the measurement you are trying to make.

Another example is shown in Fig. 1-8. A volt-ohm-milliammeter with an input resistance of 100,000 ohms cannot be used reliably to measure a voltage drop across a 5-meghohm resistor. The low resistance of the voltmeter is in parallel with the 5 meghohms, and their combined resistance is less than 100,000 ohms. That greatly reduces the voltage drop you are trying to measure.

Of course, if you know the resistance of the voltmeter, you can compute the actual voltage on the basis of the measurement taken. A much better idea, though, is to use a high-impedance voltmeter (or an oscilloscope) to make the measurement.

The manufacturer's specifications for an instrument is information that a technician should know and understand when using it to make measurements.

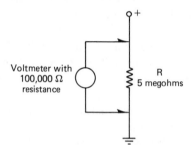

Figure 1-8 Meter loading occurs when the resistance of the meter is much lower than the resistance across which the meter is connected.

When Measurements Won't Help

Accurate measurements are very important in troubleshooting, but there are some types of troubleshooting problems where meters and oscilloscopes may not provide any useful information. As an example, an amplifier that is oscillating can produce an undesired signal in other stages. The oscillation may be originating in one amplifier stage, but you can find it in all of the other stages.

An intermittent signal is another case where making a measurement may not help.

So, to put measurements in their proper prospective, it should be said that they are useful in most troubleshooting procedures. In those cases where a measurement doesn't help, the technician must resort to other test procedures, such as those given later in this book.

Another procedure must be mentioned, one that is distasteful to many technicians: If everything else fails, it may be necessary to use the shotgun method. That means to replace each part—one at a time—until the problem is solved.

This is a last resort. However, there are a few times when none of the tricks of the trade will work. There are some cases—like complicated impedance-coupling networks—where the manufacturer suggests this approach rather than spend five days on the problem.

Measurement Efficiency

It is not productive to make a measurement that requires five pieces of test equipment to locate a problem that could just as easily be located with a voltmeter or an oscilloscope. As a rule, the more complicated the problem the more likely that very simple measurements will not be useful in locating that problem. But you have to start by eliminating the obvious. That is done by simple measurements before such techniques as sweep analysis, signal injection, and signal tracing are undertaken.

This can be summarized in a simple sentence: *Measurements for troubleshooting must be done efficiently by using the minimum test equipment necessary to locate the problem.*

SUMMARY

Troubleshooting starts by analyzing symptoms and looking for obvious defects. Reports by the user have limited value.

After these preliminary steps, the first measurement by test equipment should be the power supply voltage. If that isn't right, don't go any further until it is fixed.

There is not much said about it, but technicians often use a statistical approach to locate a defective component. Other things being equal, the component that is easiest to replace should be first on the list.

In the example given for power supply troubleshooting, the various methods of making measurements may be the key to locating the defective component.

There are two kinds of measurements: qualitative and quantitative, with some overlapping of the two. It is important to understand the type of measurement you are making. A method of qualitative analysis has been given.

Technicians must know the capabilities and the limitations of their test equipment. Use of the wrong instrument can lead to a wrong diagnosis.

Although it is very important to know how to use a wide variety of test instruments, the most efficient approach is to use the least amount of equipment necessary to get the job done.

PROGRAMMED SECTION

This section reviews the material covered in the chapter. It also introduces new material. The questions and answers are programmed to assist you. Every time you choose an answer to a question, you will be directed to a numbered block. If you answer the question incorrectly, you will be told to reread the question and choose a different answer. When you choose the correct answer, you will find a discussion and a new question.

Read the question in block 1. If you think choice A is correct, go to block 7. If you think choice B is correct, go to block 14. Continue until you reach the end of the programmed section.

---

1. The surge-limiting resistor (R_1) in a certain half-wave rectifier (like the one in Fig. 1-1) has burned out. Which of the following is correct?

   A. This could be the result of a defective electrolytic capacitor. Go to block 7.

   B. This problem cannot be caused by a defective electrolytic capacitor. Go to block 14.

---

2. You have selected the wrong answer for the question in block 17. Read the question again. Then go to the block with the correct answer.

---

3. The correct answer to the question in block 21 is B. The more likely cause of the low voltage is leakage current in C_2 or a defective (shorted) component in the power supply load circuit.

   If the voltage goes to its normal value when the load circuit is disconnected, then the trouble is in that circuit. If the output remains low, then C_2 is the likely cause.

   Resistors are not as likely to change value compared to the chance that there is a leakage current through C_2.

   Here is your next question:

   If the AC input voltage to a half-wave rectifier circuit (like the one in Fig. 1-1) is 100V, what output voltage could you expect to measure across RL?

   A. 92V. Go to block 23.

   B. 132V. Go to block 18.

---

4. You have selected the wrong answer for the question in block 16. Read the question again. Then go to the block with the correct answer.

---

5. You have selected the wrong answer for the question in block 18. Read the question again. Then go to the block with the correct answer.

---

6. You have selected the wrong answer for the question in block 19. Read the question again. Then go to the block with the correct answer.

---

7. The correct answer to the question in block 1 is A. A leaky electrolytic capacitor can cause too much current to flow through the resistor.

   The surge-limiting resistor (R_1) protects the diode during the first few cycles that occur when the capacitors are charging. This protection is needed when the circuit is first energized.

   If the capacitors have an excessive leakage current, the resistor will burn out. Remember, it may be designed to act like a fuse.

   Always check the electrolytic capacitors before replacing this resistor!

   Here is your next question:

   Can a defective diode destroy the electrolytic capacitors?

   A. Yes. Go to block 19.

   B. No. Go to block 12.

---

8. The correct answer to the question in block 20 is: Lay it on its back because the bearings in the meter are then in the position for the least amount of friction.

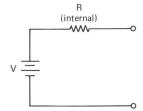

   The illustration in this block shows a battery with its internal resistance. Can you measure this resistance by connecting an ohmmeter across terminals A and B? Go to block 25.

---

9. The correct answer to the question in block 18 is A. Electrons enter the negative side of the milliammeter and leave the positive side.

   Owners of some foreign-made meters will object to this answer because their instruments may be exceptions to the rule. However, this answer is most often correct.

   Here is your next question:

   An analog VOM is connected across a 6.3V winding on a filament transformer. The meter is incorrectly set to the 50V DC scale. Which of the following is correct?

   A. The meter will display zero volts. Go to block 20.

   B. The meter will be destroyed. Go to block 15.

---

10. You have selected the wrong answer for the question in block 24. Read the question again. Then go to the block with the correct answer.

---

11. You have selected the wrong answer for the question in block 17. Read the question again. Then go to the block with the correct answer.

---

12. You have selected the wrong answer for the question in block 7. Read the question again. Then go to the block with the correct answer.

13. You have selected the wrong answer for the question in block 21. Read the question again. Then go to the block with the correct answer.

14. You have selected the wrong answer for the question in block 1. Read the question again. Then go to the block with the correct answer.

15. You have selected the wrong answer for the question in block 9. Read the question again. Then go to the block with the correct answer.

16. The correct answer to the question in block 17 is C. You can use the square wave test for a quick overview of the amplifier's frequency response. A good frequency to use for checking an audio amplifier is 2500 hertz.

    The rise time equation given in this chapter is empirical. In other words, it is based on experience rather than being derived mathematically.

    Although the rise time measurement is quantitative, there are more accurate ways to determine the bandwidth of an amplifier. Those ways are discussed in a later chapter.

    Here is your next question:

After looking for obvious defects and considering the symptoms, the first actual measurement with test equipment should be to

A. check the condition of the on/off switch. Go to block 4.

B. measure the power supply voltage. Go to block 21.

17. The correct answer to the question in block 24 is A. Reversing the diode will reverse the voltage across the electrolytic capacitor. That, in turn, will destroy the capacitors.

    Here is your next question:

A square wave test is being conducted on an amplifier. This is an example of a

A. qualitative test. Go to block 2.

B. quantitative test. Go to block 22.

C. test that can be either qualitative or quantitative. Go to block 16.

D. test that is neither qualitative nor quantitative. Go to block 11.

18. The correct answer to the question in block 3 is B. The capacitors charge to the peak voltage. Remember that the AC voltages are stated as RMS values unless otherwise stated.

$$\text{Peak voltage} = \text{RMS voltage} \times \sqrt{2}$$

$$= 100 \times 1.414$$

$$\text{Peak voltage} = 141.4\text{V}$$

There will be a voltage drop across the diode, surge-limiting resistor, and filter resistor, so the output will be somewhat lower than the calculated peak value.

Here is your next question:

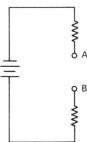

To measure current in the circuit shown in this block, the negative side of the milliammeter should be connected to

A. point A. Go to block 9.

B. point B. Go to block 5.

---

19. The correct answer to the question in block 7 is A. If the diode permits a reverse current flow there will be an AC voltage across the electrolytic capacitors. An AC voltage across an electrolytic capacitor will destroy it.

    Here is your next question:

    Refer to Fig. 1-1. Is the following statement correct? You can determine the condition of an electrolytic capacitor by measuring the DC voltage across it with an analog voltmeter.

    A. The statement is correct. Go to block 6.

    B. The statement is not correct. Go to block 24.

---

20. The correct answer to the question in block 9 is A. Assuming the meter is a Simpson 260 or its equivalent, it deflects to the average value of the input. The average value of a sine wave over a cycle is zero. So the meter displays zero volts.

    Even though the pointer on the meter is well damped, you may see some pointer vibration.

    Here is your next question:

    As a general rule, should you stand upright an analog meter that has a jeweled movement or lay it on its back when reading voltages? _____ Go to block 8.

---

21. The correct answer to the question in block 16 is B. Remember that the power supply voltage must be checked under load. In other words, check the voltage while the equipment is energized.

    NOTE

    Don't confuse the terms *load* and *load resistance*. Load is the amount of current delivered by the source of voltage. Load resistance is the amount of opposition to that current.

Here is your next question:
The output voltage is measured across the load resistance in the circuit of Fig. 1-1. It is found to be only one-fourth the required output voltage. Which of the following would be a logical step to follow?

A. Check the resistance of the filter resistor. Go to block 13.

B. Disconnect the output line of the power supply. Go to block 3.

---

22. You have selected the wrong answer for the question in block 17. Read the question again. Then go to the block with the correct answer.

---

23. You have selected the wrong answer for the question in block 3. Read the question again. Then go to the block with the correct answer.

---

24. The correct answer to the question in block 19 is B. If an analog meter is poorly damped you *might* be able to see vibration of its pointer. However, it is doubtful. In that case you wouldn't be measuring the DC voltage across the capacitor.

   The best way to determine the condition of an electrolytic capacitor is to use an ESR meter. The initials ESR stand for equivalent series resistance. It is the combination of series resistance and parallel leakage resistance of the capacitor.

   An ESR meter is very useful for determining the condition of an electrolytic capacitor. It can also be used for other types.

   Here is your next question:
   You have installed the diode backwards in the circuit of Fig. 1-1. Which of the following is correct?

   A. Parts of the circuit will be destroyed. Go to block 17.

   B. The diode will be quickly destroyed. So no other damage will result. Go to block 10.

---

25. The correct answer to the question in block 8 is: NO! Never connect an ohmmeter in a circuit with a voltage!

You have now completed the programmed section.

TEST YOUR KNOWLEDGE

1. When a measurement provides a number value, such as 3V or 27K, it is an example of a ____ measurement.

2. When a measurement does not provide a number value, but instead, provides information for making a decision, it is an example of a ____ measurement.

3. A voltmeter is connected across an on/off switch. It reads zero volts when the switch is open. Does this mean the switch is OK? ____

4. The power amplifier transistor in a certain system is shorted from the emitter to the collector. Would this make the power supply voltage higher or lower than normal? ____

5. A current meter is connected in series with the current being measured. Should the positive side of the meter go toward the positive or negative side of the circuit? ____

6. Name two kinds of voltages that can destroy an electrolytic capacitor. ____ and ____

7. An ohmmeter is being used to determine if an electrolytic capacitor is passing a leakage current. Is this a proper test? ____

8. A fuse can be checked with a voltmeter. Connect the meter ____.

   A. in parallel with the fuse.

   B. in series with the fuse.

9. After preliminary checks, the first troubleshooting measurement with test equipment should be to measure the ____.

10. A certain defective diode connects equally well in both directions. In a power supply this could result in the destruction of ____.

ANSWERS
TO TEST YOUR KNOWLEDGE

1. quantitative

2. qualitative

3. no. With the switch open, the voltmeter should display the applied voltage.

4. lower. Higher current, flowing through the transistor, increases the drop across the internal resistance of the supply. That, in turn, lowers the output voltage of the supply.

5. positive

6. AC and reverse

7. no. The 1.5V supplied by the ohmmeter is too low to be effective in testing for leakage.

8. A

9. power supply voltage

10. an electrolytic capacitor

2

Knowledge: An Important Troubleshooting Aid

CHAPTER OVERVIEW

A very important part of troubleshooting is making measurements. The procedure is to make the measurement, compare the result with what you should get, and then make a decision as to whether or not a problem exists at the point you are measuring.

This procedure cannot work, of course, unless you know exactly what you are supposed to get. In some cases, that information is given on the manufacturer's schematic, or it is given in the technical literature for the particular product you are servicing.

However, there is much information you are expected to know, and you cannot afford to look up that information each time you make a measurement. Some good examples are the polarity of the voltages on electrodes of the amplifying devices, signal configurations for those amplifying devices, and methods of connecting the devices to a DC source. Those are some of the subjects reviewed in this chapter.

This chapter is not intended to be a replacement for technical training in electronics. Rather, it is a review. Only those basic concepts that are directly related to troubleshooting are covered, and only the linear devices are discussed. Some of the thyristor and other nonlinear devices are reviewed later. Also, some of the basic digital devices are covered in Chap. 9.

A typical method of teaching about the amplifying devices used in linear circuits is to show the differences between each device. That concept is not used here. Instead, we discuss the similarities of the devices. That makes it easier to group the devices in categories.

Objectives

Some of the topics discussed in this chapter are listed here:

- What are the polarities of voltages to be expected on the various linear amplifying devices?
- What is the importance of configuration in troubleshooting?
- How are DC connections made to amplifying devices?
- How are the methods of coupling related to the types of amplifiers you are working with?
- What are the important methods of biasing linear devices?
- What are the important characteristics of power amplifiers that simplify their servicing?
- How does a superheterodyne radio serve as an example of systems used in linear troubleshooting problems?

DC VOLTAGES ON AMPLIFYING DEVICES

Figure 2-1 shows the basic amplifying devices, their related electrode terminology, and polarities of voltages to be expected on each device. These are three-terminal devices. That means that they are all similar with respect to their control terminal and input and output terminals. For that reason, they can all be represented by the rectangular boxes shown below the schematic symbols.

Note that the control electrodes are called control grid, base, or gate. Note also that all of the voltage polarities are taken with respect to the source. For example, in a bipolar transistor the voltages would be taken with respect to the emitter, and in a vacuum tube the voltages would be taken with respect to the cathode. For a field effect transistor, the source is considered to be zero volts with respect to other electrodes. These voltages are considered to be zero volts with respect to other electrodes regardless of their polarity in relation to the chassis or other components.

The NPN and PNP transistors are bipolar devices. They are different from all the other devices because they require a base *current* in order to get a collector *current*. All of the other devices require a *voltage* on their control electrodes. The arrows in all semiconductor symbols point away from a P region and toward an N region.

What actually happens inside the device is of no consideration when you are troubleshooting. If these amplifying devices are used in a conventional way, you can expect the input signal to be on the control electrode and the output signal to be either on the DC input (cathode, emitter, or source) or the DC output (plate collector or drain). Specific examples will be given later in this chapter.

In the semiconductor devices there are alternate components made by interchanging N- and P-type materials. That cannot be done in a vacuum tube. Figure 2-2 shows the alternate devices and the polarities of voltages to be expected.

Returning to Fig. 2-1, note the similarity in the polarities of voltages between the

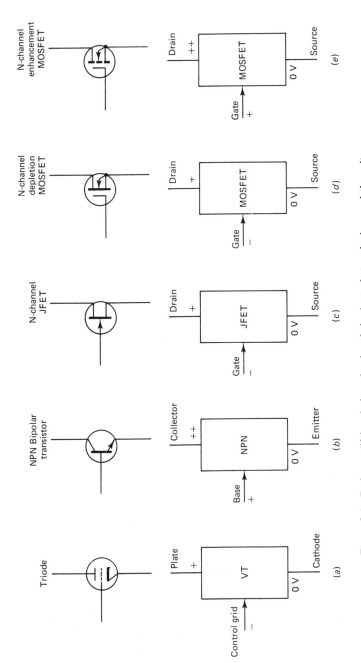

Figure 2-1 Basic amplifying devices showing their electrode terminology and the voltage expected on each electrode.

21

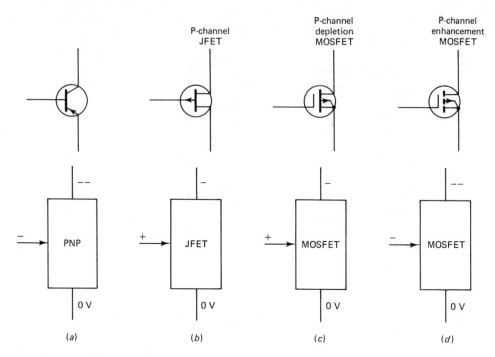

Figure 2-2 These are alternate devices that have no vacuum tube counterpart. Note the difference in symbols, especially the directions of the arrows.

bipolar (NPN) and enhancement MOSFETs (E MOSFET). Since their polarities are similar, you can expect similar methods of biasing to be used. In the same way, the vacuum tube, JFET, and MOSFET also have similar polarities, and their methods of bias are very much the same.

The enhancement MOSFET must have a forward bias before a drain current can be obtained and you should use some caution when working around these devices. Like the vacuum tubes, they can be operated at very high DC power supply voltages.

WARNING

It is not uncommon to find a voltage of 400 or 500V DC supplied to the drain of enhancement MOSFETs. The same is true of the vacuum tube.

Technicians sometimes get complacent when working around solid state equipment because such devices are normally operated at low voltages. The enhancement MOSFET is a solid state device and very often is an exception to that rule.

There are also some bipolar transistor amplifiers that are operated with a high voltage (about 100V) on the collector. These are high-frequency circuits, and the reasoning can be understood by referring to Fig. 2-3.

Recall two important basic theories about bipolar transistors. First, the base and collector junctions are reverse biased in the normal operation of a bipolar transistor in

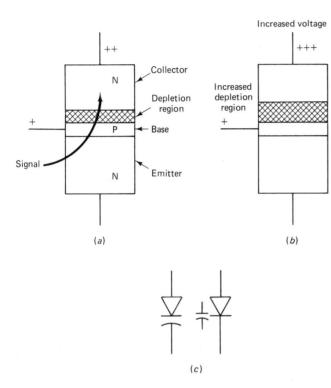

Figure 2-3 Junction capacitance in a bipolar transistor.
(a) The arrow shows how the signal sneaks through the transistor capacity without being amplified.
(b) Decreasing the capacity by increasing the reverse voltage between the base and collector.
(c) Symbols for varactor diodes. These diodes use the depletion region as an insulator.

an amplifier circuit. That means that there is a depletion region between those two sections. The depletion region is a no-man's land where no charge carriers exist. In other words, it is an insulation region. So the junction is a capacitor because it has N and P conducting materials separated by an insulator.

Recall, also, that the capacitance of a capacitor varies inversely as the distance between the plates. With a very small distance (or a thin depletion region), the capacitance between the base and collector can be quite high. A high capacitance permits high-frequency signals to sneak through the transistor without being amplified. This path is shown by the arrow in Fig. 2-3(a).

If you increase the reverse bias between the base and the collector, the depletion region is increased. This is shown in Fig. 2-3(b). Increasing the depletion region is the same as moving the plates of the capacitance apart. In other words, the capacitance is decreased.

So, in an r-f circuit, it is quite possible for the collector of the bipolar transistor to be operated at a high positive voltage in order to reduce the junction capacitance. The transistor must be designed so that it is capable of withstanding the increased reverse bias.

This discussion of the junction capacitance also applies to the varactor diode. Its symbol is also shown in Fig. 2-3. In that device the junction diode is reversed biased and it is operated as a capacitor rather than as a diode. Varactor diodes are used extensively in tuning circuits and voltage-controlled oscillators.

Figure 2-4 shows some polarities of voltages on devices that you will not see very often in your troubleshooting work. However, you should be able to make the proper measurements when they are encountered.

Go back again and look at the symbol for the triode tube in Fig. 2-1. The control electrode (or control grid) regulates the number of electrons that go from the cathode to the plate. The tube amplifies because a very small change in control grid voltage causes a relatively large change in the cathode-to-plate current.

Again, note that the control grid and plate are two metal objects separated by a vacuum and, therefore, they form a capacitor. This capacitor—as in the case of the bipolar transistor—permits the signals to sneak through from the grid to the plate without being amplified.

To eliminate the effect of this grid-to-plate capacity, an additional grid is added to the tube, as shown in Fig. 2-4(a). The screen grid shields the control grid from the plate. As a matter of fact, it acts like a Faraday screen (or Faraday shield).

The screen grid is operated with a positive voltage. That helps to draw charge carriers (electrons) away from the cathode, which, in turn, increases the plate current.

There is a problem with the tetrode of Fig. 2-4(a). Under certain conditions, high-velocity electrons striking the plate bounce off and go to the positive screen grid. That, in turn, results in a decrease of plate current.

To eliminate that condition, a third grid, called the suppressor grid, is added between the screen grid and the plate. This is shown in the pentode of Fig. 2-4(b). Usually, the suppressor grid is connected internally to the cathode in the tube. That makes it negative with respect to the plate. The negative voltage on the suppressor grid drives secondary electrons back to the plate and prevents the reduction in plate current over certain parts of the characteristic curve. If the suppressor is brought outside the tube for external use, it has a negative bias voltage.

The dual-gate MOSFET in Fig. 2-4(c) shows that there are two control electrodes. They control the motion of the charge carriers from the source to the drain.

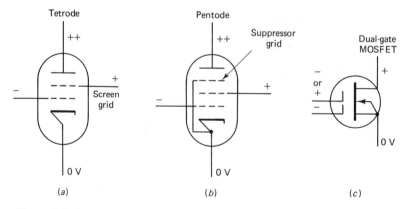

Figure 2-4 Some voltage polarities on devices that you do not often see in troubleshooting.

One of these—the one that is closest to the source—is usually used as the signal control electrode. The second gate may have a DC voltage on it to set the gain of the device, or it may be a second signal that is to be mixed (or heterodyned) with the signal on the first gate.

When you are making DC measurements around this device, remember that the second gate can be positive or negative under different conditions. It may be necessary to consult the schematic or investigate the circuit further to determine if the polarity on that electrode is correct.

All of the devices in Fig. 2-1, 2-2, and 2-4 are shown in another form in Fig. 2-5. In this case the amplifying devices are represented by models. Having a clear view of these models sometimes makes it easier to understand the operation of the devices. Let's briefly review the models.

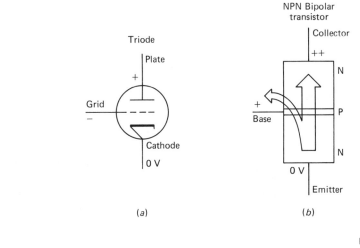

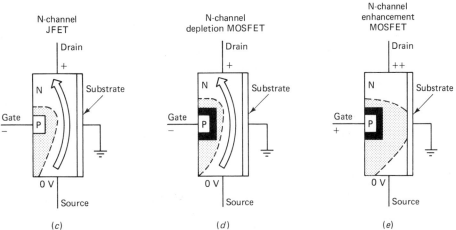

Figure 2-5 Another way of showing the amplifying devices. These illustrations are called models.

Models

The symbol for the triode tube shown in Fig. 2-5(a) also serves as a model for its operation. The cathode is heated until electrons are boiled off the surface. Those electrons are attracted to the plate.

The control grid between the cathode and the plate has a negative voltage on it. The more negative the voltage the fewer the electrons reaching the plate. Therefore, applying a negative-going signal to the control grid (which is not allowed to go positive) makes it possible to control the electron stream in accordance with the input signal.

The key to amplification by the triode is that *a small change in voltage on the grid produces a relatively large change in plate current*. When a screen grid and a suppressor grid are added to this amplifying device, they do not affect the basic operation described here. So tetrodes and pentodes are still considered to be three-terminal devices.

The NPN bipolar transistor model, illustrated in Fig. 2-5(b), shows that it is made by sandwiching a P-type material between two N-type materials. Note that the collector is more positive than the base in this transistor. In other words, the base is negative with respect to the collector.

The charge carriers are electrons. Most go from the emitter to the collector, but some go to the positive base. Increasing the base current greatly increases the number of electrons that go from the emitter to the collector.

The key amplification of this device is that *a small change in base current produces a relatively large change in collector current.*

Figure 2-5(c) shows the model for an N-channel JFET (junction field effect transistor). The large arrow represents the motion of charge carriers (electrons in this case) through the channel of the JFET. The gate, made with P material, is reversed biased with respect to the channel. Note the positive voltage on the drain and the negative voltage on the gate. With this reversed bias there is a depletion region around the gate and that limits the amount of area through which the charge carriers can pass. Increasing the negative gate voltage increases the depletion region and decreases the amount of current that can flow from source to drain. The depletion region is represented by the shaded area.

The key to the operation of this device is that *a small change in the gate voltage results in a relatively large change in drain current.* The substrate can be thought of as being the material upon which the JFET has been constructed. In a normal operation, this substrate is a semiconductor material with very little conductivity. It is usually operated at the source or the ground potential.

Figure 2-5(d) shows the model for the MOSFET. It is very much like the one for the JFET, with one exception. There is a black region around the gate that represents an insulating material. Originally these devices were called IGFETS (insulated gate field effect transistors). The insulating material is made of a metal oxide, hence the name metal oxide semiconductor field effect transistor (MOSFET).

As with the JFET, the negative voltage on the gate reverse biases the PN region and produces a depletion region. That limits the amount of current flowing from the source to the drain. The depletion region is represented by the shaded area.

Making the gate more negative decreases the drain current because of the resulting increase in the size of the depletion region.

As with the JFET, *a small change in the gate voltage causes a relatively large change in the drain current.* This device is called a *depletion* MOSFET because the operation of the gate is such that it depletes the size of the conducting region.

Figure 2-5(e) shows the model for an enhancement MOSFET. The obvious difference between this device and the depletion MOSFET is that the depletion region reaches all the way through the channel. So, when there is no positive voltage on the gate, there can be no current from the source to the drain.

In order to start current flow from the source to the drain, it is necessary to forward bias the gate. That reduces the gate-to-drain voltage and thereby reduces the size of the depletion region. In turn, that permits current to flow through the channel.

Again, observe the similarity between the voltages on the electrodes of the NPN bipolar transistor and the enhancement MOSFET. Note also the similarity and voltages on the electrodes of the vacuum tube, JFET, and depletion MOSFET.

These models are not to be thought of as methods of constructing the device. In reality, the construction is similar but not exactly like that shown in Fig. 2-5. However, by presenting the models this way it is easy to visualize what is happening in the amplifying devices.

A Resistor Model

As mentioned before, all of the amplifying devices discussed so far are three-terminal devices. That means that the signal is applied to one of the electrodes, usually the control electrode. That signal controls the number of charge carriers that pass *through* the device.

You can think of an amplifying device as being a variable resistor, as shown in Fig. 2-6. The signal on the control electrode moves the arm of the variable resistor up and down. That, in turn, controls the amount of current through the device. All of the devices shown in Fig. 2-1, 2-2, 2-3, and 2-4 operate on this principle.

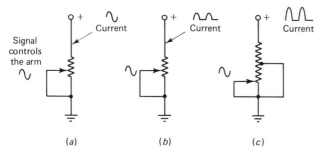

(a)   (b)   (c)

Figure 2-6 A variable resistor can be used to represent the resistance of amplifying devices.
(a) In Class A operation the input signal increases and decreases the resistance.
(b) In Class B operation the input signal can only decrease the resistance.
(c) With Class C operation a certain amount of signal is needed to get the resistance up to zero ohms.

Classes of Amplifiers

The classes of operation will also be discussed using the models in Fig. 2-6. In each case the arm of the variable resistor is shown in the zero-signal position. The signal moves the arm up and down. For the purpose of this discussion, the signal is presumed to be a pure sine wave.

The arm of the variable resistor in Fig. 2-6(a) is shown in the position for Class A operation. The incoming signal must not be strong enough to move the arm all the way to the end. If that should happen, the sine wave current through the resistor will be distorted.

In the Class B operation of Fig. 2-6(b), the resting point of the arm is at the bottom of the variable resistor. Now the applied sine wave can move the arm up, but it cannot move the arm down. Therefore, only the positive half cycles—during which the arm is moved up—will affect the current through the resistor.

As shown by the current waveform, the current increases as the arm moves up on the positive half cycle, but cannot decrease when the arm moves down. This is a representation of a Class B operation.

As you can see, the current waveform in Class B operation is not a perfect representation of the input sine wave. There is a considerable amount of distortion. However, note that the arm can now move throughout the total distance of the variable resistor on the positive half cycle. Therefore, a greater amount of control is exerted over the conducting half cycle.

Some of the amplifying devices can be operated Class C, as represented in Fig. 2-6(c). Note that zero volts is some distance above the arm and that the quiescent point is below zero volts. In order to get control of the output current, it is necessary for a certain amount of the signal to be used to get the arm up to the zero-volt point. Above the zero point, the arm can move throughout the variable resistor.

Class C operation is valuable in some circuits. For example, in a vacuum tube circuit, Class C operation is characteristic of oscillators and high-power amplifiers. It is not used in bipolar transistor circuits because in the quiescent condition the emitter-base junction would be reversed biased. As a rule, bipolar transistors are not designed for operation with a reverse bias on the emitter-base junction.

Under certain conditions, the JFETs and MOSFETs may be operated with Class C operation, but it has not been used extensively.

When the three-terminal device is operated in a conventional manner, as indicated in Fig. 2-6, the input signal is to the control electrode and the output signal at the DC output electrode is 180 degrees out of phase with that input signal. The output signal is in phase with the input signal if it is taken from the DC input electrode.

Configurations

When you remember that the amplifying device is a three-terminal device, it should be obvious that one of the terminals has to be held constant. The input signal is taken from some point to common and the output signal is taken from some point to common. Therefore, one of the three electrodes must be common. That permits three configurations of operation, as shown in Fig. 2-7.

If the input signal is to the control electrode and the output is taken from the DC output terminal, then you have a common DC input. This is sometimes called *common emitter*, *common source*, or *grounded-cathode* configuration. It is represented by Fig. 2-7(a). It has the advantage that it has the best compromise between voltage gain and power gain. Also, it gives satisfactory operation at high frequencies.

In Fig. 2-7(b), the input signal is to the DC input electrode and the output is taken from the DC output electrode. The control electrode is common. Since the control electrode is at zero volts, it acts as a shield between the input and output signals. Therefore, there is no way for the signal to sneak through the device without being amplified. This kind of configuration is called *grounded grid*, *common base*, or *common gate*. It does not have the high-voltage gain available in the conventional operation of Fig. 2-7(a). The input impedance of this circuit is low and the output impedance is high, and so it can be used as an impedance matching device. Whenever you see it, you should always consider it likely that you are working in a high-frequency circuit.

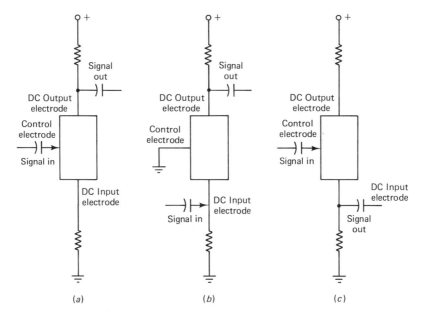

Figure 2-7   Three amplifier configurations.
(a) A conventional, or common emitter configuration (also known as common source).
(b) This configuration is called grounded grid, common base, or common gate. It is most often used for high-frequency operation. It is also used to match a low impedance to a high impedance.
(c) The follower configuration. It is used primarily for matching a high impedance to a low impedance.

Figure 2-7(c) shows a follower configuration. The input signal is on the control electrode, and the output is on the DC input electrode. This follower circuit has a high input impedance and a relatively low output impedance. For that reason it can be used as an impedance matching device. There is a voltage gain of less than 1, but it does have a power gain.

Follower circuits are used for impedance matching and also for level shifting. In direct-coupled devices, the DC level of the signal (sometimes referred to as DC offset) gets higher and higher from device to device. The follower circuit brings the DC level out of the signal down to a point where it is close to common, or zero volts.

When you are troubleshooting, you must understand what the configurations are used for and how they are affected by their DC connection. Note this very important point: These are *signal* input and output configurations. They have nothing to do with the DC power supply connections to the device. Just as one of the terminals must be common for the input and output signals, it is also true that one of the electrodes must be common for the input and output DC voltages.

DC Supply Connections

In Fig. 2-8 the common symbol is used to represent zero volts. In the conventional connection of Fig. 2-8(a), the DC input electrode is grounded and the output electrode goes to the power supply voltage. The power supply is marked positive, so it is assumed that this is one of the devices shown in Fig. 2-1. If it were one of the devices in Fig. 2-2, the power supply connection would be negative with respect to common.

In Fig. 2-8(b), the DC output electrode is common and the power supply is delivered to the DC input electrode. Note the negative supply polarity.

In the third configuration, Fig. 2-8(c), the DC input and output electrodes are delivered to separate power supplies—one positive and the other negative. This configuration is sometimes referred to as *positive-negative bias* or *long-tail bias*. Since the voltage makes the transition from negative to positive through the device, it stands to reason that the control electrode is either at, or very near to, ground potential.

The DC configuration of Fig. 2-8(c) demonstrates the importance of always considering the DC input electrode to be zero volts. That way, the polarities of the voltages on the other electrodes will still be correct.

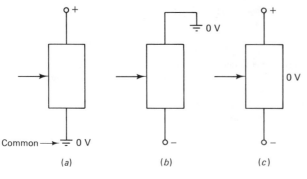

Figure 2-8 Various ways to connect the power supply to an amplifying device.
(a) The supply connected to the current output electrode.
(b) The supply connected to the current input electrode.
(c) Long-tail bias using a positive and negative supply.

The configuration you will most often see is grounded cathode, common emitter, and common source. It is important to remember the phase signals for this configuration. In Fig. 2-9, note that when the input signal is at the control electrode, the DC output signal is 180 degrees out of phase with the input. The signal at the DC input electrode is in phase with the incoming signal.

Methods of Obtaining Bias

A DC bias voltage must be applied to an amplifying device in order to get it into the quiescent condition. Figure 2-10 shows three kinds of bias. Every amplifying device can be used with these types of bias.

AGC (or AVC) bias is used primarily for receiver amplifiers. The bias voltage is obtained by rectifying and filtering the incoming signal.

Battery bias is used in portable equipment and in equipment where it is absolutely necessary to have no power supply hum and no interference on the control electrode.

In the last illustration of Fig. 2-10, a separate power supply is used for bias. This kind of bias is used extensively in high-priced, high-quality regulated power supplies. It is also used extensively in power amplifiers where a high bias voltage is needed.

AGC/AVC bias will be discussed in reference to the radio system described later

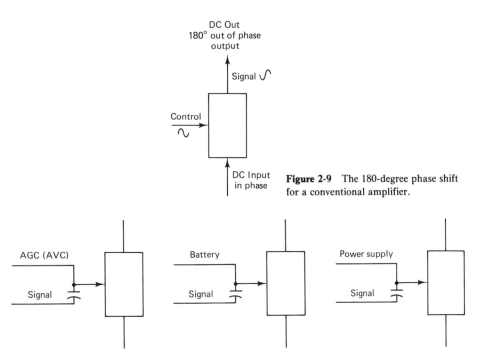

Figure 2-9 The 180-degree phase shift for a conventional amplifier.

Figure 2-10 Three kinds of bias for an amplifying device. These types can be used with all of the amplifying devices discussed so far.

in this chapter. However, there is one point that should be made at this time. A special case for AGC bias should be noted, and it is illustrated in Fig. 2-11.

There is an increase in gain in most amplifiers as the bias is moved away from cutoff. In an N-channel JFET, for example, making the bias less negative will increase the gain over a range of bias values. This is shown in Fig. 2-11(a).

For the bipolar transistor, increasing the forward bias will increase the gain up to a point. After that, the gain decreases as the forward bias increases. This is shown in Fig. 2-11(b).

Suppose, for example, that the graph of Fig. 2-11(b) is for an NPN bipolar transistor. Forward bias is accomplished by making the base positive with respect to the emitter. As the base voltage is made more and more positive, the gain increases up to a point. At some value of forward bias, the gain is maximum. After that, a further increase in forward bias results in a decrease in gain.

Figure 2-12 shows the setup for self-bias. This type of bias can be used with vacuum tubes or any of the FETs and MOSFETs, except for the enhancement type. It *cannot* be used with bipolar transistors.

Assume that it is a JFET. The current flowing into the source and out of the drain flows through source resistor R_b. That produces a voltage drop that makes the source positive with respect to common. The gate electrode is connected to common through R_a.

Normally, there is no current flowing in this gate circuit, so the gate is essentially at ground potential. Since the gate is essentially at ground potential, and the source is positive with respect to ground, it follows that the gate is negative with respect to the source. That is the same as saying the source is positive with respect to the gate.

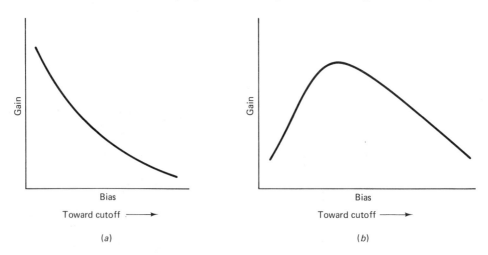

Figure 2-11   Two types of AGC bias.

(a) For conventional AGC bias, the gain increases as the bias decreases. In this illustration the gain is decreasing with an increase in bias.

(b) With bipolar transistors, there is a certain point of bias for maximum gain. Any change in bias from that point will result in a decrease in gain.

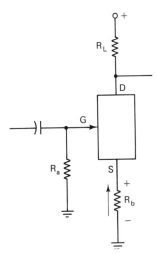

Figure 2-12 The setup for self-bias. This type of bias is used with tubes and field effect transistors.

This type of bias is called self-bias, or automatic bias, or source bias. When used with a vacuum tube, it is called cathode bias.

When there is a resistor in the emitter of a bipolar transistor, it is not used for bias. Its correct name is *temperature stabilizing resistor* because it helps to stabilize the bipolar transistor amplifier against changes in temperature.

Bipolar transistors and enhancement MOSFETs operate with similar DC voltage polarities. So they are used with similar kinds of bias. A popular way to get this bias is shown in Fig. 2-13.

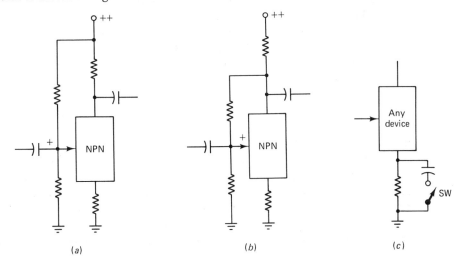

Figure 2-13 Types of bias for bipolar transistors and enhancement MOSFETs.
(a) A simple voltage divider bias.
(b) Voltage divider bias with negative feedback.
(c) Another way of getting negative feedback. In this case an open switch results in negative current feedback for the device.

In Fig. 2-13(a), a voltage divider is used between the positive power supply voltage and common. That voltage divider forward biases the base of the bipolar transistor (or enhancement MOSFET). Voltage divider bias is the type you will most often encounter for these devices.

There is a trade-off that is characteristic of all amplifiers and should be understood by a troubleshooting technician. That trade-off is between the amount of gain of an amplifier and the range of frequencies it can pass. In other words, amplifier bandwidth and gain are trade-offs.

Anything you do to increase the gain will automatically decrease the bandwidth. Conversely, anything you do to decrease the gain will automatically increase the bandwidth.

A good way to remember this is if you decrease the gain of the amplifier down to the point where the output and input signals are the same, then the amplifier has the same gain as a piece of straight wire and has a very, very wide bandwidth.

One way of decreasing the gain of a bipolar amplifier is to connect the bias divider circuit directly to the collector, as shown in Fig. 2-13(b). For an enhancement MOSFET, it would be connected to the drain. Now the voltage divider is taken from a point where the output signal is 180 degrees out of phase with the input signal. Part of the output signal is fed back to the input by the voltage divider in Fig. 2-13(b). That out-of-phase signal partially cancels the input signal. The result is a decreased gain and a resulting increased bandwidth.

There is another way to achieve negative feedback. In some amplifiers, the signal is fed back through a capacitor. That isolates the input and output DC voltages but permits the 180 degree out-of-phase signal to be returned to the input to cancel part of the input signal.

Another common way of decreasing the gain is illustrated in Fig. 2-13(c). A resistor at the DC input electrode is used. All of the amplifying devices can be operated with this resistor at the DC input position. The signal at that input is in phase with the incoming signal.

So, when the incoming signal goes positive, the voltage drop across the resistor increases, causing the DC input electrode to follow the signal. The DC electrode follows the signal up and down and decreases the gain of the amplifier. That, in turn, increases its bandwidth.

If the switch in Fig. 2-13(c) is closed, a capacitor is connected in parallel with the resistor. That capacitor charges and discharges and maintains the voltage across the resistor at a constant value. Now the input voltage at the DC input electrode is held constant. That increases the gain of the amplifier, but its bandwidth is decreased.

A simple bias circuit for bipolar transistors is shown in Fig. 2-14. This circuit is relatively unstable compared to voltage divider bias. It is used in low-cost bipolar transistor amplifiers.

The emitter resistor is used for stabilizing against temperature changes. Without it, a thermal runaway condition can occur. Thermal runaway is an upward-spiraling condition in which an increase in a transistor's temperature causes its current to increase, and an increase in its current causes its temperature to increase. It rapidly destroys the transistor.

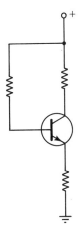

Figure 2-14 A simple bias circuit. This circuit is used in toys and very low-cost equipment. It is not the best way to bias a bipolar transistor.

COUPLED CIRCUITS

There are four methods of coupling the signal from one amplifier to the other. They are shown in Fig. 2-15. Coupled circuits can be the clue to the kind of operation that you are troubleshooting.

The R-C coupling shown in Fig. 2-15(a) is very common in low-frequency amplifiers. There is no tuned circuit, so you are not dealing with an r-f or i-f amplifier. The

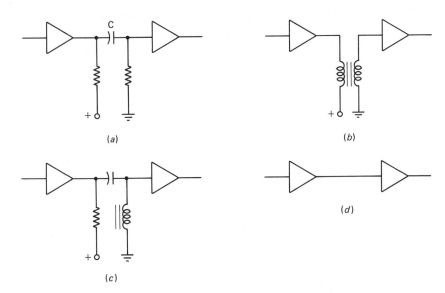

Figure 2-15 Four methods of coupling signals between circuits.
(a) R-C coupling.
(b) Transformer coupling.
(c) Impedance coupling.
(d) Direct coupling.

35

disadvantage of this kind of coupling is that very low frequencies see a high reactance in the coupling capacitor (C_2). So the gain falls off rapidly at low frequencies.

Transformer coupling, shown in Fig. 2-15(b), is used throughout a wide range of amplifier frequencies. If the transformer is untuned or if it has an iron core, as shown, you are definitely dealing with a low-frequency amplifier, usually in the audio range. On the other hand, if the transformer is tuned, you are dealing with a high-frequency r-f amplifier.

One version of impedance coupling is shown in Fig. 2-15(c). In modern circuits impedance coupling has become quite complex with computer-designed configurations.

The overall purpose of the circuit is to pass a band of frequencies. In that sense, it is similar to transformer coupling. In the circuit shown, the inductor is used to increase the high-frequency response. That, in turn, increases the range of signals passed through the two amplifiers. Impedance coupling is made with combinations of resistors, capacitors, and inductors. If it is an r-f circuit the capacitors and/or the inductors will be tunable in order to set the range of frequencies that can be passed.

Figure 2-15(d) shows direct coupling. The signal does not pass through any reactive components as it goes from the output of the first amplifier to the input of the second. This gives it a wide frequency range, but the system is not without its problems, as discussed in the next section.

Level Shifting

Direct coupling requires a good power supply because the DC levels of the second amplifier must be higher than the DC levels of the first one. That causes a condition known as level shifting and it is illustrated in Fig. 2-16. Note that the input DC voltage

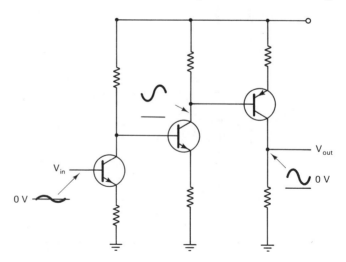

Figure 2-16   The problem of level shifting in direct-coupled amplifiers is partially reduced by using the PNP (upside-down) configuration.

of the second amplifier has to be at the same DC level as the output level of the first amplifier. Therefore, the collector voltage of the second amplifier must be even higher.

It is not unusual to have four or five direct-coupled amplifiers in order to get a desired gain at a broad band width. The problem is that the DC level of the signal is shifted and the power supply must be able to accommodate the increased DC levels in each stage.

The relationship between the signal and the DC level is shown in Fig. 2-16. This signal offset is undesirable. It is partially compensated for by the upside-down PNP transistor as the last stage. This PNP transistor returns the signal level to very near common. At the same time, it produces additional gain in the chain of amplifiers.

Instead of the upside-down transistor, an NPN transistor emitter follower transistor circuit could be used. It will produce a broad bandwidth but has the disadvantage that its voltage gain is less than 1. Therefore, an amplifier is being used without any additional gain. That may be undesirable when gain is an important factor in the design.

Differential Amplifiers

A differential amplifier is shown in Fig. 2-17. One way to use this circuit is to deliver two signals—180 degrees out of phase—to inputs x and y. The output signal is proportional to the difference between the input signals. Since one signal is going positive while the other is going negative, their difference is greater than either signal taken by itself. That explains the high gain of the differential amplifier.

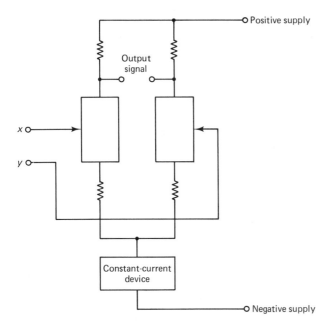

Figure 2-17 A differential amplifier has a very high gain. This type of amplifier is often used as the input circuit for integrated circuit amplifiers.

If inputs x and y are connected together, it is called a common-mode connection. A signal delivered to the common-mode input should produce no output signal because the voltages at the output terminals rise and fall together. One way of testing differential amplifiers, then, is to look for zero-volt output signal when the input is common mode.

An important feature of this amplifier is the constant-current device that controls the amount of current into the two amplifiers. Because of this device, anything done to increase the current in one of the amplifiers will automatically decrease the current by the same amount in the other.

Differential amplifiers are often used in operational amplifiers. In Fig. 2-17, long-tail bias is used, but some of the new operational amplifiers can be used with a single-ended power supply.

Power Amplifiers

There are some important characteristics of power amplifiers to watch for when troubleshooting these circuits. They must have a high input signal voltage, but the output signal voltage is usually not high. A power amplifier is used to convert signal voltage into signal current, and it is the current that produces the power in the output circuit.

You can expect a very low voltage gain in a power amplifier. They are often operated with heat sinks with cooling fins. Except for the high currents involved, a single-ended power amplifier looks very much like a voltage amplifier.

The output of a power amplifier operates a device that requires high current, often some form of transducer or motor. A loudspeaker is an example of a transducer.

Figure 2-18 shows a single-ended power amplifier. The input signal is direct coupled from the previous audio voltage amplifier stage. The emitter stage is unbypassed for a broader frequency response. This transistor is normally mounted on a heat sink.

The circuit is inefficient compared to other power amplifiers because the transistor is operated Class A. However, it is very popular in low-priced equipment.

Figure 2-19 shows a push-pull power amplifier. This design will produce more power than a single-ended amplifier. It is designed in such a way that when the signal

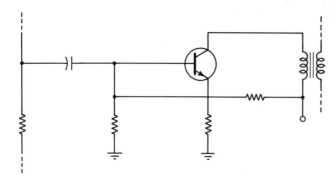

Figure 2-18 A single-ended power amplifier. This is not an efficient circuit.

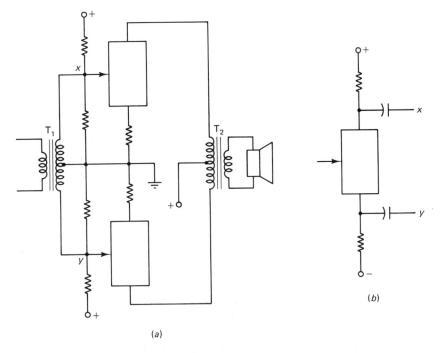

(a)

(b)

Figure 2-19 Power amplifiers in a push-pull configuration produce more output power than the single-ended type.
(a) A phase splitter is needed. In this case a transformer is used.
(b) Phase splitting can be accomplished by an amplifier, as shown here.

at x is going positive the signal at y is going negative. The output is produced by first one amplifier and then the other to complete a cycle.

Because of the way they are designed, these amplifiers must be operated Class AB, B, or C. In an audio amplifier, that is usually Class B or AB. Class AB amplifiers have a slight forward bias. It is not biased at cutoff like Class B amplifiers described earlier in this chapter. The slight forward bias is used to overcome the 0.7V present between the emitter and base of bipolar transistors.

Without the slight forward bias, there is a period of time when neither amplifier is conducting. That produces an undesirable condition in the output called *crossover distortion*, illustrated in Fig. 2-20. So, if the amplifier in Fig. 2-19(a) is made with bipolar transistors, you can look for some kind of forward bias on the transistors to keep them slightly forward biased.

If one of the two amplifiers and the push-pull circuit doesn't conduct, the other one may still produce a very distorted output.

The 180-degree out-of-phase signals for the push-pull amplifier of Fig. 2-19(a) is produced by a transformer that is center tapped. The out-of-phase signals may also be produced by a phase splitter, as shown in Fig. 2-19(b). Here, one output is taken from the DC output terminal and one is taken from the DC input terminal. That makes the

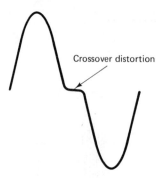

Figure 2-20   Crossover distortion is
primarily a problem with bipolar transistor power amplifiers.

signals at *x* and *y* 180 degrees out of phase, and those out-of-phase signals can be used
to operate the two amplifiers. At one time, only vacuum tubes or bipolar transistors
were used in the push-pull amplifier configuration. However, special MOSFETs,
called VFETs, are capable of amplification power and they are now showing up as
power amplifiers in various circuits. These VFETs operate the same as any other
MOSFETs, but they are designed to pass relatively high currents.

   A popular power amplifier that does not require a phase splitter or output transformer is shown in Fig. 2-21. It is presented in a simplified form, but the essential
parts are present.

   A PNP and NPN transistor are used to alternately charge and discharge capacitor C. On the positive half cycle the NPN transistor is forward biased and the capacitor
charges in the direction shown by the solid arrows. On the next half cycle the negative-
going input signal cuts off the NPN transistor and forward biases the PNP type. That
transistor conducts heavily and discharges the capacitor in a path shown by broken
arrows.

   Since the current through the speaker is in one direction during charging and the
opposite direction during discharging, the speaker is actually carrying an amplified
version of the AC signal delivered to this amplifier.

   This kind of amplifier is very popular in operational amplifier output circuits

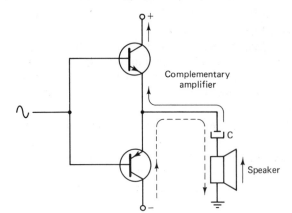

Figure 2-21   This is a simplified version
of a complementary amplifier. Note that
no phase splitter is required.

and in many receiver systems. It does not need to be operated by polar transistors. Power-operating VFETs can also be used, but a vacuum tube cannot be used in this configuration because there is no such thing as a complementary vacuum tube amplifier. They all operate with positive voltages on the plate with respect to the emitter.

Long-tail bias is shown in the circuit of Fig. 2-21, but single-ended versions are available.

THE RECEIVER AS AN ELECTRONIC SYSTEM

As mentioned in the preface, a simple AM receiver will be used to represent a system. This is not because this book is about repairing radios but, rather, because this simple system is easy to understand and is familiar to most technicians.

The basic idea of the receiver is reviewed in block diagram form shown in Fig. 2-22.

All receivers have four sections in common:

- There must be an antenna system to deliver the incoming signal to the selection part.
- There must be a tuner that permits the receiver to select one station and reject all others.
- There must be some kind of detector to separate the carrier wave and audio waveforms. The audio waveform is then delivered to the audio amplifier shown in the block diagram of Fig. 2-22.

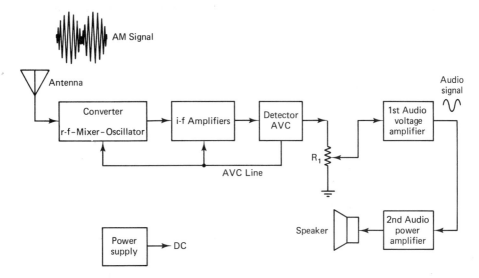

Figure 2-22 Block diagram of a simple AM radio. This system will be used throughout the text to explain system concepts.

- Finally, there must be some method of reproducing the signal. It is the speaker in the system of Fig. 2-22.

Instead of using a converter—that is, a combination of an r-f amplifier, a mixer, and an oscillator—it is possible to have these in separate stages, as shown in Fig. 2-23. In this case, the combination results in a separate mixer stage.

In the circuits of Fig. 2-22 and 2-23, it is necessary to use a nonlinear stage to combine the signals.

IMPORTANT

You cannot heterodyne or modulate signals in a linear amplifier circuit. Nonlinear means that the amplifier is operated Class B or Class C or, in the case of the biopolar transistor versions, Class AB. Linear amplifiers are operated Class A, so their output signal is the same shape as the input signal.

The oscillator frequency is tuned so that it is 455 kilohertz above the incoming signal. When the radio is tuned from station to station, the oscillator frequency is also tuned. The 455-kHz i-f signal, which is always the difference between the r-f and oscillator signal, is delivered to the i-f amplifiers. The i-f amplifiers provide a high amount of gain for the signal and their output signal is delivered to the detector.

The detector separates the audio from the i-f signal. The audio signal is delivered to the audio amplifier stages through the volume control (marked R_1 in the block diagram). Note the AVC line that comes from the detector. This AVC line carries a DC voltage. The stronger the incoming signal, the higher the AVC voltage and the more negative feedback that occurs in the amplifier. This AVC voltage controls the gain of the first stages in such a way that a stronger signal results in less gain.

The output of the volume control goes first to an audio voltage amplifier. Its purpose is to increase the signal voltage to a sufficient amplitude for driving the power amplifier. That power amplifier is a second audio amplifier in the receiver. Its purpose is to convert the high-voltage input signal amplitude to current variations that can operate the transducer (speaker).

All of the stages (with the possible exception of the detector) are fed by a DC power supply voltage.

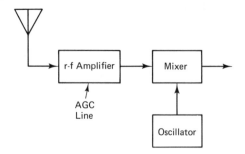

Figure 2-23 The front end of a receiver may consist of individual sections. When these sections are combined into one, the circuit is called a converter.

The block diagram of Fig. 2-22 represents the receiver shown in schematic form in Fig. 2-24.

The following lists the purpose of each component in the schematic of Fig. 2-23:

RESISTORS

- R_1 and R_2—voltage divider bias for Q_1
- R_3—emitter stabilization resistor for Q_1
- R_4, R_6, and R_7—the components in this voltage divider circuit have several purposes
- R_5—emitter stabilization resistor for Q_2
- R_7—volume control
- R_8 and R_9—voltage divider for biasing Q_4
- R_{10} and R_{11}—voltage divider bias for the push-pull amplifier; this circuit provides Class AB operation
- R_{12}—emitter stabilization resistor for the push-pull amplifier
- R_{13}—acts to drop the supply voltage to a lower value for the voltage amplifiers; it is also part of a decoupling filter composed of R_{13} and C_8

CAPACITORS

- C_1—couples the r-f signal to the base of Q_2 and, at the same time, prevents the base bias of that transistor from being grounded through windings on T_1 and T_2
- C_2—emitter-resistor bypass capacitor; it increases the gain of Q_1 by eliminating signal voltages on the emitter of that transistor
- C_3—tuning capacitor for the primary of T_3
- C_{4a} and C_{4b}—r-f and oscillator tuning capacitors
- C_5—feedback capacitor to increase the high-frequency gain of Q_2; note that it is connected to the primary of T_4 in such a way that it provides a regenerative feedback
- C_6—tuning capacitor for the primary of T_4
- C_7—emitter-resistor bypass capacitor used to prevent degeneration and thereby increase the gain of Q_2
- C_8—part of the voltage decoupling filter (see also R_{13})
- C_9 and C_{10}—along with R_6, this is the AGC (AVC) filter
- C_{11}—coupling capacitor between the volume control and the audio voltage amplifier
- C_{12}—this small capacitor removes any residual i-f signal that gets through the system
- C_{13}—false bass tone control; high audio frequencies are eliminated so as to give the output a more bass sound

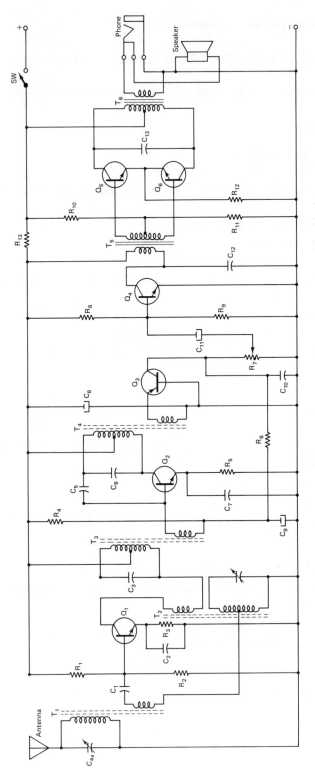

Figure 2-24 Circuit for an AM radio. The purpose of every component is explained in the text.

TRANSFORMERS

- T_1—r-f transformer that is part of the circuit that selects one of many stations
- T_2—oscillator transformer; it provides tuning for the oscillator signal and regenerative feedback for the oscillator signal.
- T_3—first i-f transformer
- T_4—second i-f transformer
- T_5—phase splitter for the audio push-pull amplifier
- T_6—output transformer for the push-pull audio amplifier

TRANSISTORS

- Q_1—converter
- Q_2—i-f amplifier
- Q_3—detector; this transistor has no forward bias, so it is being operated Class B.
- Q_4—audio voltage amplifier
- Q_5 and Q_6—push-pull audio amplifiers

FINDING TROUBLE IN A DEAD RECEIVER

A good way to start troubleshooting any system is to divide the system in half. Using the example of a radio again, the volume control is an easy place to start.

If you inject an audio-frequency signal at the top of the volume control you should be able to hear the audio at the speaker. (You may have to adjust the volume control for sound.)

If you hear the sound you can assume the circuits between the volume control and output are OK. If not, move the test signal forward, point by point, until you hear the sound. When you do, you have just passed the point where the trouble is located.

If you do hear the sound when you inject the signal at the volume control, then it follows that the trouble is somewhere between the volume control and the antenna.

You have to change to a *modulated* r-f signal. In an AM radio you would use 455 kHz for the i-f frequencies and any frequency from 550 to 1600 kHz at the antenna. Trace through the radio from the volume control to the antenna using the proper signal.

The preceding method works very well. You can add the use of an oscilloscope to observe the waveforms as they pass through various sections. You can use a sweep generator to look at the bandpass of the i-f transformers and amplifiers.

That is good experience, but, it is a good way to go broke. This is *not* saying that test equipment is not important and useful.

Tapping the center tap of the volume control with a screwdriver is a quick way to inject a signal. The tapping should produce a clicking noise in the speaker. If it does, then the audio system is probably OK and your time would be better spent looking at the sections in front of the volume control. After looking for obvious faults and mea-suring the power supply voltage, a few quick checks should be used to determine where the fault lies.

In Chap. 3 and 4 a number of quick tests will be discussed and the idea of divid-ing the system in half will be mentioned again.

OPERATIONAL AMPLIFIERS

In mathematics an *operator* is a symbol that tells you what to do. For example, a plus sign indicates that you are to add the second number to the first. Likewise, a division sign means that you are to divide the first number by the second.

Operational amplifiers were used in early analog computers because they were used to perform arithmetic operations. A long time ago it was learned that it was unnecessary to design a new amplifier for every new arithmetic operation. *If the gain of an amplifier is sufficiently high, its characteristics are almost completely controlled by the type of feedback that is used with that amplifier.*

Figure 2-25 shows two typical operational amplifier circuits. Power supply con-nections are not shown with the op amp symbols. The + and − signs refer to the signal noninverting and inverting input terminals. Note that a signal delivered to the inverting input results in an output signal that is 180 degrees out of phase with the input.

There are some important characteristics that you should know about this type of amplifier in terms of troubleshooting. The difference in voltage between the input terminals is maintained at zero volts. The amplifier will do anything that it has to in order to keep that DC voltage at zero volts.

If the feedback resistor (R_f) is large, then the amplifier must have a high gain in order to overcome the feedback signal opposition. Remember that *the feedback signal*

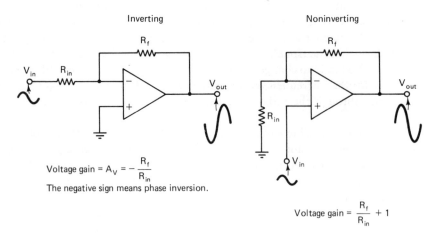

Figure 2-25   Two examples of operational amplifiers: inverting and noninverting.

is used to maintain the input DC voltage at zero volts. If the feedback resistor is small, then the gain is low because not much output signal is needed to feed back to the common node input in order to maintain the voltage at zero volts.

Operational amplifiers found a new lease on life when they became integrated circuits. That allowed them to be used for a very wide variety of applications. You will see operational amplifiers in every type of electronic system today.

Op amps are highly reliable and require little troubleshooting. They are typically low-frequency amplifiers, so an oscilloscope check of the input and output signals is the usual procedure. Look for distortion, especially clipping, in the output signals.

For many years the op amps used only long-tail (positive/negative) bias. Now there are single-ended op amps.

As with any integrated circuit (IC) device, check the power supply inputs.

The next statement is controversial so use your own judgment: When you check any integrated circuit for signals or voltages, make your measurements at the pins, *not* at the printed circuit board connection.

Figure 2-26 shows the way to probe an IC. The reason some publications and schools take a stand against this is because it is easy to slip off the terminal. If the

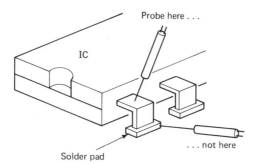

Probe here . . .

IC

. . . not here

Solder pad

Figure 2-26 In the opinion of the author this is the correct way to probe signals and voltages on an integrated circuit.

probe shorts between two pins, you can cause another problem that you didn't have before. However, if you check at the board instead of the pin, you still don't know if the signal or voltage is actually being delivered to the IC.

THE OHMMETER TEST FOR TRANSISTORS

When a transistor is defective in a circuit voltage, measurements can usually be used to determine that it is bad. Furthermore, there are in situ testers that test the transistor while it is wired into place. In situ testers usually apply a low operating voltage to the transistor and operate it as an oscillator. The theory is that if it won't operate as an oscillator the transistor must be bad.

Remember that an oscillator is nothing more than a regenerative feedback amplifier circuit that has a very high gain and a very narrow bandwidth.

When you are replacing a transistor, it is desirable to test the new one before it is soldered into place. The very best testers are those that check the beta of the transistor. Remember that the beta of a transistor is determined as follows:

$$\beta_{DC} = \frac{IC}{IB} \ (I_E \text{ held constant})$$

$$\beta_{AC} = \frac{\Delta IC}{\Delta IB} \ (I_E \text{ held constant})$$

Ohmmeters can be used to check bipolar transistors and other components. However, as a general rule, the ohmmeter tests are not as reliable as beta tests or dynamic operating tests. Nevertheless, you may be in a situation where you do not have a beta checker available. Certainly the ohmmeter test is better than nothing.

The ohmmeter test is illustrated in Fig. 2-27. In order to perform this test, it would be best if you know which of your ohmmeter leads is positive and which is negative. In the illustration, it is assumed that the lead with the alligator clip is the negative side of the ohmmeter. Remember that an ohmmeter supplies a voltage to the circuit being tested.

You should *never* test a transistor, or any other semiconductor device, when the ohmmeter is on the ×1 (times one) position. In that position the ohmmeter is capable of supplying high current that can destroy a semiconductor device. So, for this test, it is assumed that the ohmmeter is *not* on the ×1 position. (The ×100 scale is recommended.)

In Fig. 2-27(a), the negative lead of the transistor is connected to the base. The positive lead is then connected first to the collector and then to the emitter. If you are testing a PNP transistor, the connection in (a) will forward bias both of the junctions. If it is an NPN transistor, it will cause a reverse bias. So the transistor can be identified as being either an NPN or PNP type.

Suppose it is the PNP type. Then both of the connections in Fig. 2-27(a) should show a relatively low resistance. On the other hand, if it is an NPN type, both of the connections should show a high resistance.

Changing to the positive lead on the base, as shown in Fig. 2-27(b), should reverse the conditions that you got in (a). So, if you got low resistance in both directions in (a), you should get a high resistance in both directions in (b). On the other hand, if you got high resistances in (a), you should get low resistances in (b).

You have checked the emitter-base and collector-base junctions to see if they are capable of rectifying.

A third test should also be used. It is illustrated in Fig. 2-27(c). Now the ohmmeter is connected from emitter to collector. You should measure a high resistance in both connections shown. The base lead is open, so the transistor is not forward biased. That, in turn, means a high emitter-collector current. This particular ohmmeter test is essential for power transistors that sometimes have a shorted emitter-to-collector connection.

Some technicians prefer the test illustrated in Fig. 2-28. It is easy to make a

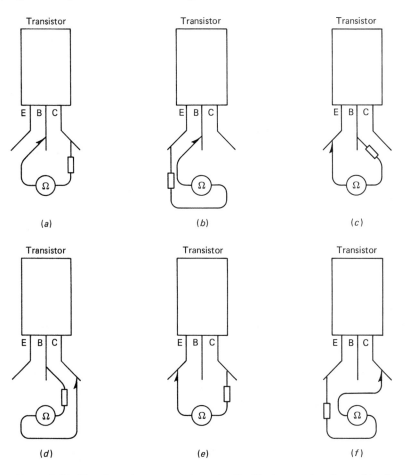

Figure 2-27　Use of an ohmmeter to determine if a bipolar transistor is capable of performing its function.

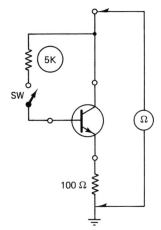

Figure 2-28　A better way to check a bipolar transistor with an ohmmeter.

simplified test jig where the transistor can be plugged in and the ohmmeter used for the power supply.

With the switch open, the ohmmeter should show a high resistance. When the switch is closed, the transistor is forward biased and the transistor should have a low resistance. This test is better than the ohmmeter test because it is dynamic—that is, it tests the transistor's ability to control current flow by changing the amount of base current. However, it is still not as good as the beta test.

In any case, the judgment about which test is better depends on the preference of the technician and the types of troubleshooting the technician is doing.

SUMMARY

There are some basic voltages and polarities that a technician must know when troubleshooting. Technicians do not have time to look up these voltages on schematic diagrams or in books when they are searching out the cause of a trouble. It is something that they must *know* in order to troubleshoot efficiently. These voltages and polarities have been summarized in this chapter. There has also been some discussion on basic circuit configurations that often require troubleshooting.

A logical question at this time is: How is this troubleshooting information affected by the use of integrated circuits in some modern systems? The answer is that, with integrated circuits, troubleshooting is almost entirely on a system basis. So the techniques of signal tracing and signal injection are most important for those systems. Always keep in mind, however, that even integrated circuit systems have outboard transistor amplifiers and transistors used for specialized purposes. You will need the information in this chapter for troubleshooting those types of circuits.

Because it provides a basis for troubleshooting discussions, a radio system has been reviewed briefly, as well as the purpose of each component in that radio.

PROGRAMMED SECTION

Instructions for using this programmed section are given in Chap. 1.

1. The amplifier shown in this block

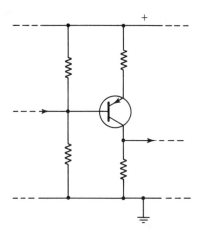

    A. has a gain less than 1. Go to block 7.
    B. can be used for level shifting. Go to block 14.

2. The correct answer to the question in block 23 is B. If R_2 opens, there will be no conduction of transistor Q_1 through R_2. That will cause the base voltage of Q_2 to float to a high positive value of voltage. The resulting high base current through Q_2 can destroy that transistor.
    Here is your next question:
Which of the following types of solid state amplifying devices might have $+450$V on its DC output electrode?
    A. depletion MOSFET. Go to block 25.
    B. enhancement MOSFET. Go to block 15.

3. You have selected the wrong answer for the question in block 22. Read the question again. Then go to the block with the correct answer.

4. The correct answer to the question in block 20 is B. It is not possible to bias a bipolar transistor with self-bias.
    Here is your next question:
Can an enhancement MOSFET amplifier be self-biased with a source resistor?
    _____ Go to block 29.

5. You have selected the wrong answer for the question in block 10. Read the question again. Then go to the block with the correct answer.

---

6. You have selected the wrong answer for the question in block 15. Read the question again. Then go to the block with the correct answer.

---

7. You have selected the wrong answer for the question in block 1. Read the question again. Then go to the block with the correct answer.

---

8. The correct answer to the question in block 21 is B. Any amplifier that is designed to produce mixing or heterodyning must be nonlinear. Your understanding of this concept is important for understanding tests for intermodulation distortion.

Here is your next question:

Refer again to Fig. 2-24. If R_9 is open, then transistor Q_4 will be

A. saturated. Go to block 23.

B. cut off. Go to block 19.

---

9. The correct answer to the question in block 22 is A. Voltage gain in the power amplifiers is not important. At zero signal volts, it would take about 0.7V between the emitter and base to start either transistor into conduction. The part of the signal on the base that is between 0V and 0.7V would be lost. That, in turn, would result in crossover distortion. The slight forward bias on the bases—that is, the Class AB operation—keeps both transistor bases biased to the point of conduction and eliminates the crossover distortion.

Here is your next question:

Refer to the system in Fig. 2-24. What is the purpose of the circuit composed of C_9, R_6, and C_{10}?

A. It is a filter. Go to block 21.

B. It is a circuit for coupling a signal from one point to another. Go to block 16.

---

10. The correct answer to the question in block 26 is B. If you missed this question, you should make it a point to review *all* of the voltages on amplifying devices.

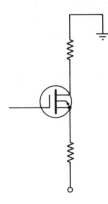

Here is your next question:

For the connection shown in this block the gate bias voltage should be

A. positive with respect to common. Go to block 5.

B. negative with respect to common. Go to block 28.

---

11. You have selected the wrong answer for the question in block 23. Read the question again. Then go to the block with the correct answer.

---

12. You have selected the wrong answer for the question in block 21. Read the question again. Then go to the block with the correct answer.

---

13. You have selected the wrong answer for the question in block 20. Read the question again. Then go to the block with the correct answer.

---

14. The correct answer to the question in block 1 is B. This circuit is often used at the output of direct-coupled amplifiers.

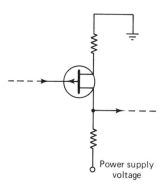

Power supply voltage

Here is your next question:

The power supply voltage for the circuit in this block should be

A. positive. Go to block 26.

B. negative. Go to block 24.

---

15. The correct answer to the question in block 2 is B. Be careful when working in enhancement MOSFET circuits.

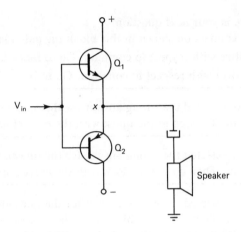

Here is your next question:

Consider the simplified drawing of a complementary circuit shown in the illustration in this block. Point x is accidentally shorted to common. This will

A. destroy Q_1. Go to block 6.

B. destroy Q_2. Go to block 27.

C. not destroy either transistor. Go to block 20.

---

16. You have selected the wrong answer for the question in block 9. Read the question again. Then go to the block with the correct answer.

---

17. You have selected the wrong answer for the question in block 26. Read the question again. Then go to the block with the correct answer.

---

18. You have selected the wrong answer for the question in block 28. Read the question again. Then go to the block with the correct answer.

---

19. You have selected the wrong answer for the question in block 8. Read the question again. Then go to the block with the correct answer.

---

20. The correct answer to the question in block 15 is C. With long-tail bias the voltage at point x is very near zero volts. So shorting the point to common will not affect the condition of Q_1 and Q_2.

Here is your next question:

Which of the following best describes the purpose of R_3 in the system of Fig. 2-24?

A. It is used for self-biasing Q_1. Go to block 13.

B. It is used for stabilizing the amplifier against temperature changes. Go to block 4.

---

21. The correct answer to the question in block 9 is A. This filter eliminates audio variations from the detector and delivers a DC voltage as part of the bias for Q_2.

This is AVC bias. Transistor Q_2 also gets some simple forward bias through resistor R_3. It is not unusual for two different types of bias to be used with an amplifier.

Here is your next question:

A mixer stage in a receiver should be

A. Class A. Go to block 12.

B. Class B. Go to block 8.

---

22. The correct answer to the question in block 28 is A. The output of Q_1 is 180 degrees out of phase with the signal at V_{in}. That output is inverted again, so the output of Q_2 is *in phase* with the signal at V_{in}. Then Q_3 inverts the signal one more time. That makes the output of Q_3 (at V_{out}) 180 degrees out of phase with the input signal.

Note that Q_3 is in a common emitter configuration. That follows because its input signal is at its base, and its output signal is at its collector. The DC connections have nothing to do with its signal configuration.

Here is your next question:

In the radio circuit of Fig. 2-24, the push-pull amplifiers (Q_5 and Q_6) are operated Class AB. This is accomplished with a slight forward bias on their bases. The forward bias is obtained with a voltage divider composed of R_{10} and R_{11}. Why is the Class AB operation needed in this section?

A. to eliminate crossover distortion. Go to block 9.

B. to provide a higher voltage gain. Go to block 3.

---

23. The correct answer to the question in block 8 is A. Resistor R_{10} acts to (among other things) pull down the DC bias voltage. If that resistor is open, there won't be enough voltage drop across R_8. That, in turn, will allow the positive voltage on the base to float to a higher value. The transistor will start to conduct very heavily into saturation.

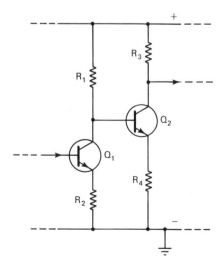

Here is your next question:

Consider the direct-coupled amplifiers shown in this block. If resistor R_2 opens,

A. transistor Q_2 will not be affected. Go to block 11.

B. transistor Q_2 will likely be destroyed. Go to block 2.

---

24. You have selected the wrong answer for the question in block 14. Read the question again. Then go to the block with the correct answer.

---

25. You have selected the wrong answer for the question in block 2. Read the question again. Then go to the block with the correct answer.

---

26. The correct answer to the question in block 14 is A. This is a P-channel JFET, so its drain must be negative with respect to its source. That is another way of saying that its source must be positive with respect to its drain.

Here is your next question:

Which type of field effect transistor is biased like a PNP transistor?

A. N-channel depletion MOSFET. Go to block 17.

B. P-channel enhancement MOSFET. Go to block 10.

---

27. You have selected the wrong answer for the question in block 15. Read the question again. Then go to the block with the correct answer.

---

28. The correct answer to the question in block 10 is B. In this configuration the DC source voltage must be negative because the drain must be positive with respect to the source. The gate must always be negative with respect to the source. So it must be negative with respect to common.

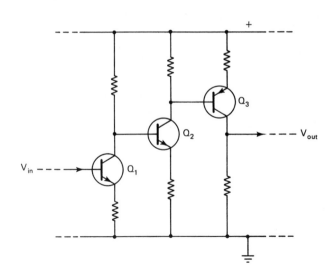

Here is your next question:
In the direct-coupled amplifier circuit shown in this block, the signal at V_{out} is

A. 180 degrees out of phase with the signal at V_{in}. Go to block 22.

B. in phase with the signal at V_{in}. Go to block 18.

---

29. The correct answer to the question in block 4 is no.

You have now completed the programmed section.

TEST YOUR KNOWLEDGE

1. In a Class A amplifier made with an NPN transistor you would expect the DC base voltage to be _____.
   A. positive with respect to the collector.
   B. negative with respect to the collector.
2. Which types of field effect transistors have the same polarity of voltage on their electrodes as vacuum tubes? _____ and _____
3. Increasing the gain of an amplifier will automatically _____.
   A. increase its bandwidth.
   B. decrease its bandwidth.
4. What is the first measurement made with test equipment when troubleshooting? _____
5. What type of diode is used as a capacitor? _____.
6. To reduce base-collector capacity, the collector voltage is _____.
   A. increased with respect to the base voltage.
   B. decreased with respect to the base voltage.
7. In a tetrode vacuum tube the screen grid voltage should be _____.
   A. positive with respect to the cathode.
   B. negative with respect to the cathode.
8. Which of the following amplifier configurations has a low input impedance and a high output impedance? _____
   A. common gate
   B. common drain
9. Which of the amplifying devices discussed in this chapter can be used with either AGC or power supply bias? _____
10. Which of the following is used to increase the bandwidth of an amplifier? _____
    A. regenerative feedback
    B. degenerative feedback

ANSWERS
TO TEST YOUR KNOWLEDGE

1. B
2. N-channel JFET and N-channel depletion MOSFET
3. B
4. The power supply voltage
5. varactor (sometimes called by its trade name, Varicap)
6. A
7. A
8. A
9. any of the devices
10. B (regenerative feedback is used with oscillators)

3

Troubleshooting

with a Meter

CHAPTER OVERVIEW

A certain amount of personal opinion is involved in troubleshooting methods. One technician may prefer to use a voltmeter for most troubleshooting problems; another will always reach for the oscilloscope leads.

Even though you may have your own preference, you should be familiar with all of the methods of troubleshooting. This chapter deals with the voltmeter as a trouble-shooting instrument.

Both analog and digital meters are available. Although you may use only one type, you should know the advantage and disadvantages of each. Also, you should know their limitations.

Here are a few precautions regarding the use of voltmeters:

- Don't measure across a high-impedance circuit with a meter that has a low input resistance.
- Don't try to measure nonsinusoidal voltages. Most meters are calibrated to read only sine wave voltages.
- Don't use a meter to measure integrated circuits unless you *know* the probes are static-free.
- Don't use a voltmeter to measure the open circuit voltage across a cell or battery because it won't tell you anything useful about its condition.
- Don't use a lot of test equipment for measurements when a simple voltmeter can do the job.

Some of these ideas are expanded in the chapter.

Objectives

Some of the topics discussed in this chapter are listed.here:

- How is AC current measured with a voltmeter?
- How is the input resistance of an analog volt-ohm-milliammeter estimated?
- How to you make use of a meter turnover?
- How do you get maximum accuracy with voltmeter measurements?
- What are some methods of testing transistors and thyristors with a VOM?
- How do you avoid misuse of the dB scale on an analog meter?
- What is a method of estimating voltages?
- How do you know when a voltmeter test won't work?

ABOUT THE METER

In addition to their ability to measure voltage, current, and resistance, there are some important features of meters used in modern servicing.

In analog meters there is a choice between jeweled and taut-band meter movements. The most important difference is the way the return force is applied to the pointer. This return force is needed to return the pointer to zero and to provide the necessary opposition as the pointer moves upscale. Figure 3-1 shows the concepts.

The taut-band movement is the more rugged of the two and will hold its accuracy over a longer period of time. More will be said about these two movements in a later discussion.

Some analog meters have a mirror on the scale to prevent parallax. Parallax error is a problem in measurement accuracy. It occurs when you view the pointer at an angle.

An example of parallax is when a passenger looks at the speedometer pointer from an angle. Since the passenger is sighting along the pointer and speed scale at an angle, it *appears* that the speed of the car is lower than it really is.

A mirror on the scale can eliminate the error. The procedure is to view the meter with one eye. When you are looking at it in a position where you cannot see the reflection of the pointer in the mirror, then you are looking at it from a vertical (most accurate) position.

The low-power ohms scale on meters is needed to make resistance measurements

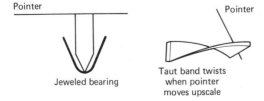

Jeweled bearing

Taut band twists when pointer moves upscale

Figure 3-1 The basic idea behind jeweled and taut-band meter movements.

in semiconductor circuits. As shown in Fig. 3-2, the ohmmeter leads have a voltage difference that can turn a semiconductor junction on. In this case, that would put R_1 in parallel with R_2 and the junction resistance.

If you don't have a low-power ohm scale, you can estimate the parallel resistance of R_1 and R_2 and disregard the resistance of the forward-biased transistor. Then the measurement in Fig. 3-2 will tell you if either resistor is open or greatly changed in value.

You could, of course, reverse the leads of the ohmmeter to prevent the junction from being forward biased to the point where it conducts. That is *not* necessarily the best solution.

You have to consider speed to be a very important factor when measuring with a meter, scope, or any other measuring device. The fastest technicians clamp the common lead of the meter onto a circuit common point and make all measurements with a single probe. That eliminates a lot of wasted time that would occur when you move both probes around the circuit.

If you measure across R_2 in the setup of Fig. 3-2 with the ohmmeter, the junction is not turned on, so you know whether R_2 is the correct value. Then move the probe to the top of R_1 and estimate the parallel resistance of R_1 and R_2:

$$(R_1 \times R_2) \div (R_1 + R_2) \qquad \text{(disregard junction R)}$$

You can tell if R_1 is far out of tolerance by this method.

Use of the low-power ohms scale is preferable if you have that provision on your VOM. It does not supply enough voltage to turn the junction on. So R_1 in Fig. 3-2 is no longer in parallel with R_2 and the resistance can be measured directly.

Analog meters are still very useful whenever a high input impedance is not absolutely necessary. One advantage of the analog meter scale is that it is easy to use in applications where adjustments require a null or a peak.

Table 3-1 is a comparison of analog and digital meters. You will want to add your own opinion based on the types of measurements you make in your work.

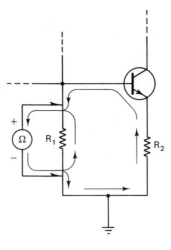

Figure 3-2 In this measurement the ohmmeter is forward biasing the emitter-base junction of the transistor. That produces a parallel path for ohmmeter current through resistors R_1 and R_2. Obviously, the ohmmeter will not correctly display the resistance of R_1.

TABLE 3-1   COMPARISON OF DIGITAL AND ANALOG MULTIMETERS

Digital Multimeters

A $3\frac{1}{2}$ digit or $4\frac{1}{2}$ digit meter usually is adequate. Choose one that has the greatest number of counts on each range, as this increases the accuracy and resolution.

Microprocessor-controlled meters often offer additional features, such as autoranging, and a variety of measurement functions like temperature and capacity.

Particularly useful when troubleshooting linear ICs, they are even more useful when used in conjunction with an oscilloscope.

Unless they are adequately shielded, they usually are susceptible to interference from extraneous noise sources.

DVMs offer 10 MEG or more of input impedance.

Their price ranges from $40 to $700 for general-purpose troubleshooting.

Analog Multimeters

They are subject to the destruction of meter movement because of excessive current flow through the circuitry. Choose one that offers a lot of input protection.

Usually the least susceptible to interference from extraneous noise sources. However, they can malfunction around an r-f transmitter.

Because of their lower input impedance and consequent ability to inhibit noise pickup, they are often the best choice for servicing appliances.

Incorporating multiple scales, they make several related readings available simultaneously.

Their built-in electromechanical integration smooths jittery readings.

They vary in price from $10 to $400.

Reprinted with permission from the September 1987 issue of *Electronic Servicing & Technology.* Copyright 1987, Intertec Publishing Corp., Overland Park, Kansas.

If you are using an older VOM, make sure that there are no static charges on the probes because those charges can destroy some types of integrated circuits (especially CMOS).

Vacuum tube voltmeters are still being sold. They are analog meters with a very high input resistance. There is nothing wrong with using a vacuum tube voltmeter, assuming there are no static charges on the probes.

Measuring AC Current

Traditionally, analog meters do not have AC current-measuring capability. That does not mean you can't measure current with them, but it requires a special test setup.

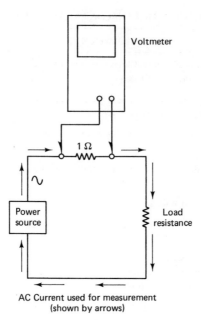

Voltmeter

1 Ω

Power
source

Load
resistance

AC Current used for measurement
(shown by arrows)

Figure 3-3   A 1-ohm resistor can be used
to measure AC current. This illustration
shows the procedure.

Figure 3-3 shows how it is done. A 1-ohm resistor is connected in series with the circuit current to be measured. The voltmeter measures the voltage across that resistor. The current is numerically equal to the voltage when the inserted resistance equals 1 ohm:

$$I = \frac{V}{R} = \frac{V}{1}$$

so

$$I = V$$

If you insert a 10-ohm resistor, the voltage reading must be divided by 10.

As long as the resistance of the inserted resistor is less than 10 percent of the total resistance, the method just described will give a good estimate of the current.

Ohms per Volt

Analog meter movements are often rated according to their sensitivity stated as the number of ohms-per-volt. A frequently missed question in the CET test is given here:

Question

A certain meter is rated at 10,000 ohms per volt. When it is being used to measure 5V, its resistance is

$$10,000 \; \frac{\text{ohms}}{\text{volt}} \times 5 \; \text{volts} = 50,000 \text{ ohms}$$

This is

A. correct.

B. not correct.

Answer

B

You *can* determine the approximate resistance of the meter by using the ohms-per-volt rating and the maximum (full-scale) voltage on a specific voltage range. Figure 3-4 shows the calculation.

So you can tell how much resistance the meter is offering when making a voltage measurement. For all practical purposes, that resistance is the same as the resistance of the multiplier. (The meter movement resistance is disregarded.)

Example

A certain meter with a 0–10V scale and a meter sensitivity of 20,000 ohms per volt is being used to measure a circuit voltage of 3V. The meter resistance is 60 ohms. How much resistance does the meter offer? In other words, what is the input resistance to the meter?

Solution

Use the approximate method shown in Fig. 3-4:

$$\text{Meter resistance} \approx \text{max voltage} \times \text{ohms per volt}$$

$$\approx 10 \times 20,000$$

$$\text{Meter resistance} \approx 200,000 \text{ ohms}$$

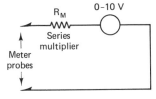

Approximate meter resistance = ohms per volt × full-scale voltage.

Figure 3-4 The approximate meter resistance can be obtained by this simple method.

The meter resistance for a given scale is the same regardless of the particular voltage being measured. So whether you are measuring 3V or 8V on the 10V scale, the measurement has nothing to do with the meter resistance on that scale.

Meter Loading

When a voltmeter is connected across a resistor for making a voltage reading, there is always a certain amount of *meter loading*. That means the meter reduces the circuit resistance and the *measured* voltage will be less than the actual voltage without the meter.

If the meter resistance is at least ten times the resistance of the resistor it is connected across, you can assume that the measurement is a good approximation.

Calibration

It is useless to troubleshoot with a multimeter if that meter is not accurate. It is important that you periodically check the accuracy of your meters. This can be done in a number of different ways.

To check the accuracy of your voltmeter, connect it in parallel with one that is known to be good. This is shown in Fig. 3-5.

If you have an adjustable power supply, it is useful to check the meter scale at a number of different voltage values.

For a quick check, connect the meter across a $1^{1}/_{2}$V carbon-zinc dry cell (La Clanche type) that is known to be good. That is a reasonably accurate 1.5V source.

Another place to get an accurate voltage is in the +5V supply for CMOS integrated circuits. That voltage must be maintained accurately. For AC voltage measurements you can use a 6.3V filament transformer (or transformer with a higher secondary voltage).

Current-reading meters are checked by connecting them in series, as shown in Fig. 3-6. Since the same current flows through both meters, they should indicate the same value.

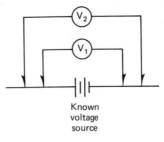

Figure 3-5 This is the method used for checking the accuracy of a voltmeter against one known to be accurate.

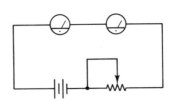

Figure 3-6 This is the method of checking the accuracy of a current meter against one that is known to be correct.

A precision resistor is a good way to check the resistance ranges of an ohmmeter. Use 1 percent resistors for this check.

Most voltmeters can be used for measuring *only* sine wave AC voltages. The AC scale is calibrated to read RMS value of a sine wave voltage. In the case of an analog VOM with a permanent magnet, moving coil* movement, the meter deflection is actually proportional to the *average value* of the voltage being measured. However, the scale is *calibrated* in RMS volts. This is based on the simple relationship:

$$V = V(\text{ave}) \times 1.1$$

where V is the RMS value and 1.1 is the form factor of a sine wave.

Remember, that relationship is only true for pure sine wave voltages. So the scale reading is incorrect if you are measuring a nonsinusoidal voltage.

You can use an AC meter calibrated for RMS value to measure the average value of a nonsinusoidal waveform if you know the form factor of the waveform. By definition

$$\text{Form factor} = \frac{\text{RMS value}}{\text{Average value}}$$

so

$$\text{Average value} = \frac{\text{RMS value (shown on meter)}}{\text{Form factor}}$$

Try this trick with your VOM on a square wave. A square wave has an RMS value equal to its average value. So its form factor is 1.

The meter may have input capacitance and/or inductance that changes the waveform slightly. So when the waveform is nonsinusoidal, the traditional way of making a measurement is to use an oscilloscope to measure the peak-to-peak value. That is the value you will see most often on schematics.

There is another trick used with VOMs to determine the nature of the nonsinusoidal waveform being measured. This trick employs the use of meter turnover. The procedure is to measure an AC voltage with the probes in one direction, as shown in Fig. 3-7. Then reverse the probes (see Fig. 3-7). If you get the same reading both ways, you know that the full-cycle average value of the waveform is zero.

An example of when the turnover measurement can be useful is in setting zero offset on a function generator. Figure 3-8 shows how the offset adjustment affects the signal.

The offset adjustment doesn't show the exact position of the signal. If you apply a signal that doesn't have zero offset to a differential amplifier—like the amplifiers you find at the input of an operational amplifier—the DC level will cause an unbalance. That, in turn, will usually result in distortion of the output signal.

---

*In popular usage, the term *D'Arsonval* is often applied to this type of meter movement. The original D'Arsonval movement was quite different.

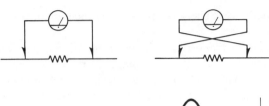

Figure 3-7 This procedure is used to ensure that the full-cycle average of an AC waveform is zero.

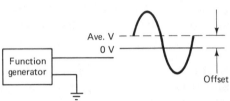

Figure 3-8 Function generators often have a DC offset adjustment. The waveform indicates that DC offset is present. The AC meter measurement demonstrates the DC offset.

Use the turnover measurement just described and adjust the generator offset until there is no turnover.

These quick tests will give you an idea if the meter is working properly, but they are not sufficiently accurate for calibrating a meter.

Taut-Band and Jeweled Meters

Taut-band instruments have two very important advantages. One is that they are more rugged than the jeweled type. Also, there is no problem with stiction.

In the jewel type, a jewel bearing is used to hold the pointer and a spiral spring is used to provide the restoring force. The problem is that there is some amount of starting friction sometimes (called *stiction*) between the pointer and the cup that it rides on.

Think of stiction as being a *sticking friction*. When a very slight change is required the jewel, especially in older instruments, tends to stick until a sufficient force is applied to start it. This is why you are advised to tap the meter gently when using the meter to make very small changes in adjustments.

With a taut-band meter there is no stiction problem.

To reduce the problem of stiction in a meter with a jeweled movement, *the meter should always be put on its back so that the pointer is centered in the cup.* Note in Fig. 3-9 that when the meter is standing the pointer rides down on the cup. This is especially true for older meters.

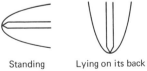

Standing   Lying on its back

Figure 3-9 The accuracy of measurement for a jeweled meter movement is affected by the position of the meter during measurement. Note that when the meter is standing the pivot point is moved down to the side of the jeweled bearing.

Accuracy of Voltmeter Measurement

Meter manufacturers state that maximum accuracy occurs when the meter is reading full scale.

Suppose, for example, the meter is being used to measure on the 10V scale. The manufacturer says that is will give a 2 percent accurate reading on that scale. That means it will be 2 percent accurate *if you are reading 10 volts.* Usually when the measurement is less than full scale the accuracy decreases.

That is why you are told to read meters when they are deflecting to some value in the upper one-third of the range for best accuracy.

Meter Probes

In addition to direct probes, there are two important types of probes used with voltmeters: high-voltage probes and demodulator probes.

High-voltage probes act as a voltage divider to drop most of the high voltage being measured. This type of probe is shown in Fig. 3-10. If the probe is designed specifically for the meter, you simply connect it into the voltage terminal of the meter and read the voltage by using the indicated multiplying factor.

If you are designing your own series multiplier, be sure to use a full-scale voltage value for the highest voltage range. Determine the current that flows through the meter using the reciprocal of the ohms-per-volt scale. Then calculate your series resistor so that it can be used for a voltage at least $1\frac{1}{3}$ times the value of the highest voltage that you expect to measure with this multiplier. Figure 3-11 shows a typical high-voltage probe.

A demodulator probe is shown in Fig. 3-12. This probe is useful for making voltage measurements in very-high-frequency circuits. For example, if you want to measure the voltage in a television i-f stage, your meter may not be able to handle the 41-mHz measurement. However, with a demodulator probe you are removing the r-f and looking only at the waveform envelope. Your meter will surely be able to give you some kind of indication.

This is very important: Unless the modulation is a pure sine wave you will get an indication, but it will not be an accurate measurement of the modulated waveform RMS value.

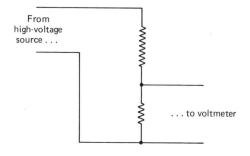

From high-voltage source . . .

. . . to voltmeter

Figure 3-10   There is nothing more complicated in a high-voltage probe than a simple resistor voltage divider.

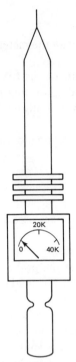

Figure 3-11 High-voltage probes are made many different ways. This is an example of a self-contained high-voltage probe.

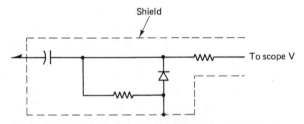

Figure 3-12 The demodulator probe shown here is easily constructed. It is very useful for measurements in r-f circuits.

Sometimes you are simply using the meter to see if there is an AC signal voltage present. Make sure there is a series capacitor so that DC in the circuit will not produce an upscale reading when you are checking only for AC.

BATTERY TERMINAL VOLTAGE

This is a good place to repeat the first step in trouble-shooting:

Step 1: After preliminary visual tests and symptom analysis, *measure the power supply voltage.*

Almost any voltmeter will present a sufficiently high resistance so that it does not draw enough current to properly load a battery for testing.

Figure 3-13 shows a voltmeter being used to measure the terminal voltage of a dry cell. Assume that this dry cell is very old and has a high internal resistance. Under

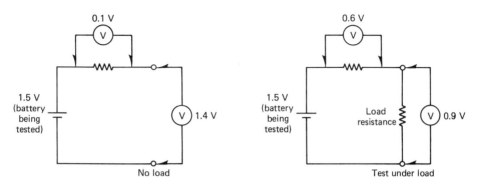

Figure 3-13 This illustration shows why the condition of a dry cell cannot be determined by an open terminal measurement. The measurement must be made with the cell under load, as shown in the right-hand drawing.

normal load the terminal voltage drops to a value that is too low to make the battery useful. However, measuring the terminal voltage with a voltmeter produces such a low value of voltage drop that the terminal voltage seems to indicate that the battery is good.

So a battery should always be tested under load. If you are troubleshooting an electronic system that uses battery power, the best load is to use the circuit the battery is supposed to supply power for.

Measure the battery terminal voltage with the equipment turned on. Reject a battery that has an under-load terminal voltage less than 80 percent of the full rated voltage. (The 80 percent value is average. It may differ from specific applications.)

Ohmmeter Testing

CAUTION

Never use the ×1 scale on an ohmmeter when you are checking a pn junction. Some meters will supply junction-destructive currents on the ×1 scale!

An obvious use of the ohmmeter scale on a DMM is for measuring the resistance of resistors. Some early digital multimeters had poor accuracy on the ohms scale. This has been taken care of in the newer designs and, in fact, they can now be used for very accurate measurements.

Figure 3-14 shows how an ohmmeter can be used to evaluate the condition and type of bipolar transistor. This isn't a test that you would run very many times before you get tired of it. If you do a lot of transistor testing, it is better to build a beta checker, or use the beta checker that is available on many of the new digital multimeters.

Two ohmmeter tests for vacuum tubes are shown in Fig. 3-15. The tests shown are open or closed filament and the filament-to-cathode shorts.

Other tests show whether electrodes are directly shorted. For example, in a high

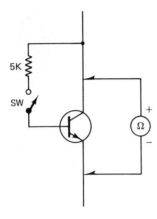

The ohmmeter resistance reading should be high with SW open
and lower with SW closed. Reverse ohmmeter for PNP test.

Figure 3-14 This ohmmeter check of a
transistor was shown in a previous chapter. You can easily make a test setup for
this measurement.

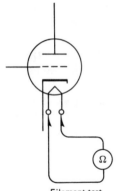

Filament test

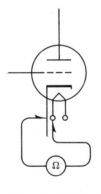

Filament-cathode test

Figure 3-15 Here are two important
ohmmeter checks for vacuum tubes. A
third possible check is to measure the
resistance between the grid and cathode.

amplification factor tube the grid might become shorted to the cathode. As with the
transistor test, this is a once-in-a-while quick test. Tube testers do a much better job.

Field effect transistors can be tested with an ohmmeter, as shown in Fig. 3-16.
Commercial testers do a much better job of testing.

The ohmmeter test for thyristors are shown in Fig. 3-17.

Testing Capacitors with a VOM

The condition of capacitors is difficult to determine with an ohmmeter. Usually, an
ohmmeter does not deliver more than a 1½Volt output. So when you connect the
leads across a capacitor it may not tell you that the capacitor is leaking with its normal
operating voltage. You can, of course, tell that it is not shorted, but, for very small
capacitors, it is difficult to tell if the capacitor is open.

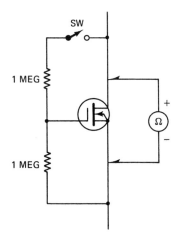

Ohmmeter should show decrease in resistance when switch is closed.

Figure 3-16 The MOSFET check shown can be used for other types of field effect transistors.

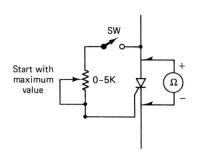

When SCR is first connected (SW open), the ohmmeter shows high resistance. When SW is momentarily closed, the resistance drops and remains at the lower value.

Figure 3-17 Thyristors can be checked using this technique. You should always start with the resistor adjusted to its highest value.

You can easily rig a time-constant test for small capacitors, as shown in Fig. 3-18. The capacitor should charge to the supply voltage in a time (seconds) equal to the resistance (ohms) multiplied by the capacity (farads):

$$T = R \times C$$

Make the resistance high enough so that you can observe the capacitor charging on the upward swing of the pointer. Remember that an open capacitor would also result in a voltmeter measurement of the supply voltage.

A number of years ago a technician from Florida sent a test that was used for checking leakage of electrolytics. This is passed on to the reader the way he told it. (See Fig. 3-19) It may be useful when it is necessary for you to test frequently electrolytic capacitors. The measured value would vary for different circuits. Note that the voltmeter is being used as a current measuring device in this application.

This test can be used any time you want to measure leakage currents, and it is a better test than trying to use the ohmmeter voltage.

Technicians often check electrolytic capacitors by bridging them with a capacitor known to be good. This test is illustrated in Fig. 3-20. The capacitor is bridged with the equipment energized.

You should not use this test in a semiconductor circuit. What you *should* do is make sure that the equipment is first turned off, connect the capacitor in using convenient alligator clips, and then turn the equipment on.

Transients caused by momentary bridging of capacitors when the equipment is on can destroy semiconductor devices.

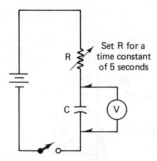

Figure 3-18 Capacity can be checked with this simple test setup. If the time constant is long, the time taken for the capacitor to charge to about 63 percent of the supply voltage can be measured accurately. Then $C = T \div R$.

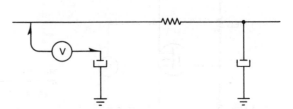

Figure 3-19 This is an unusual method of checking electrolytic capacitors.

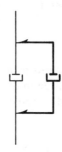

Figure 3-20 Bridging is a popular method of checking electrolytics. However, it is not a good method because it can destroy semiconductor devices in the circuit.

Diode Test with an Ohmmeter

Ohmmeters are sometimes used to check inductive circuits to determine whether or not there is continuity. That test can be misleading. For one thing, it does not show whether there are shorted turns in the equipment.

Keep this very important fact in mind: An ohmmeter delivers *voltage and current*. That current can turn on a transistor, or a diode, or other active device. If there is a high inductance in the circuit, the inductive kickback when removing the ohmmeter can easily be sufficient to destroy a semiconductor device.

Estimating Voltages

Voltmeter measurements are important for evaluating components and circuits, but they are useless if you don't know what the voltage is supposed to be. One way to determine the required voltage is to consult schematics. Another useful way is to estimate what the voltage should be.

Consider the voltage divider shown in Fig. 3-21. What voltage should the voltmeter indicate? Even without a schematic, it should be obvious that the amount of voltage across the resistor (R_2) is determined by the proportion of that resistance value to the total resistance in the circuit. Disregarding the very small current from the transistor base, the voltage at point A can be estimated easily as follows:

$$V_B = V_{bb} \left(\frac{R_2}{R_1 + R_2} \right)$$

With a little practice, you can estimate these voltages very quickly.

The forward voltage from the emitter to the base is also very important. It is obtained by measuring the emitter voltage and subtracting from the base voltage, as shown in Fig. 3-22.

Why not simply measure the emitter-base voltage directly? The answer is that you should practice to reduce the amount of time you spend locating circuit troubles.

If you start moving both probes around the circuit you are going to waste valuable time. The fastest technicians report that they clip the negative lead to common and make all measurements with the remaining probe. This procedure is especially fast with a DVM that has autoranging.

Amplifier Voltmeter Test

You don't need a schematic drawing to check whether most amplifiers are in working order. There are several basic measurements that are very useful.

In a Class A amplifier the collector (or plate, or drain) voltage should be about one-half the power supply voltage. Therefore, after measuring the power supply volt-

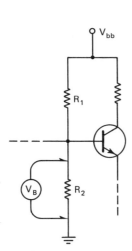

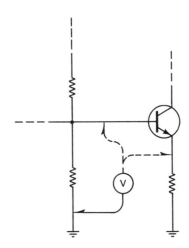

Figure 3-21 You can
readily estimate voltages
in voltage dividers.

Figure 3-22 This method of measuring emitter-to-base voltage is preferred
because the meter is permanently
connected to the common point.

age, your next measurement is at the collector (or other) output. That will tell you if the amplifier is working and if it is conducting an amount typical of a Class A amplifier.

If the voltage is too high it means the amplifier is not conducting hard enough. That can be due to insufficient bias or a poor transistor.

If the voltage is too low it means the transistor is conducting too hard. That can also be a bias problem or a bad transistor.

Once you have located the faulty stage, further tests can be used to determine which component is at fault.

The test just given isn't sufficient to warrant the removal of a transistor, but it is a good place to start. You may also determine the condition of a low-frequency amplifier by measuring the AC input signal and AC output signal. Most amplifiers should show a certain amount of gain—that is, the output AC voltage should be more than the input.

You can use an AC voltmeter for this test, but you must know the frequency limit of your meter. Usually, a voltmeter can be used at all audio frequencies.

Your voltmeter may not be sensitive enough to measure the input signal strength. In that case, a prescaler can be used. This is an amplifier with a specific voltage gain. It amplifies the weak input signal so the voltmeter can measure it.

As shown in Fig. 3-23, you have to divide the measurement with the voltage gain of a prescaler.

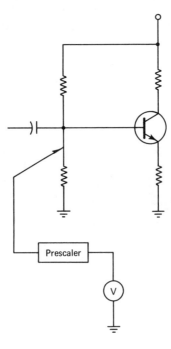

Figure 3-23 Prescalers can be used for measuring unusual voltages. The prescaler is connected in series with the voltmeter, as shown here.

NOTE

The term *prescaler* is also used for a dividing circuit that lowers a frequency. That allows a frequency counter to measure a frequency that is above its normal range.

In a power amplifier it is quite possible that the amplifier is working very well but the output voltage is not a high as the input voltage. This is typical because the output of a power amplifier should be a signal *current*. Where that current flows through a low value of load resistance, such as in a speaker or transformer, the voltage drop may be relatively small even though the power gain is high.

So, when you are making an AC input/output voltmeter test you need to know what kind of an amplifier you are measuring in. Very often you can tell that simply by the physical size of the transistor. Power transistors are usually large compared to voltage amplifiers and they run at a higher temperature.

Short-Circuit Voltmeter Test

The short-circuit test shown in Fig. 3-24 for *bipolar transistor amplifiers* is used by technicians to determine if the transistor is not only passing current but has control over that current.

Think again about measuring the collector voltage. Suppose in a particular case the collector voltage is half the supply voltage, but it is due to the fact that there is an emitter collector short circuit (or partial short circuit). The transistor is bad even though the collector voltage measurement is correct.

By shorting the emitter to the base, as shown in Fig. 3-24, the transistor will be turned off. The collector voltage should go high. In fact, it should be almost equal to the power supply voltage.

There are two cases where the short-circuit test is useless. The first is a class B

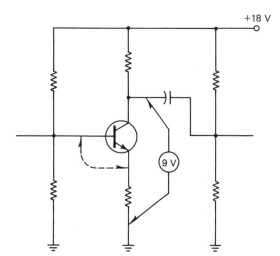

Figure 3-24   This illustration shows how the useful emitter-base short test is made.

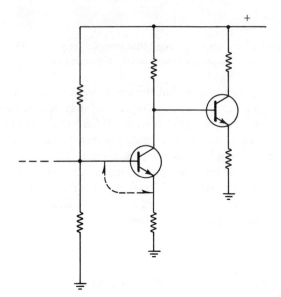

Figure 3-25 In this case the emitter-base short circuit can destroy the second transistor.

amplifier. This type of amplifier is biased at cutoff and will not conduct unless there is an input signal. Since the amplifier is already at cutoff, the short circuit will not produce any change in the DC output voltage.

The second case where this test is useless, and in fact, *destructive* is shown in Fig. 3-25. Here the transistors are direct coupled. The base voltage of the second transistor is equal to the collector voltage of the first.

If you connect an emitter-base short circuit on the first transistor, its collector voltage will rise to a very high value. That high value on the base of the second transistor *will almost surely destroy it!*

SUMMARY

Troubleshooting requires tests and measurements. A technician's skill at troubleshooting is directly dependent on the proper use of test equipment when making the required tests and measurements.

The use of voltmeters, ohmmeters, and current meters was the subject of this chapter.

Analog and digital meters are used in servicing, but the digital meters are preferred. One reason is their high accuracy.

Another reason is that they are easy to use. With autoranging it is not necessary to worry about which scale to use. It is not necessary to interpret an answer. With an analog meter you have to interpret the position of the pointer between the marks. You also have to mulitply the indicated value by some multiplier. These steps are not required with an autoranging meter.

A good technician takes time to check the calibration of the meters used for troubleshooting. It isn't enough just to make measurements for troubleshooting information. The information must be reliable.

Reliable information doesn't help unless the *required* value is known. The troubleshooting procedure requires that *measured* values be compared with *required* values.

Some methods of determining the required values were discussed in the chapter. The required voltages for amplifiers are an example.

A taut-band meter movement is rugged and able to maintain accuracy. The restoring force for the pointer is a twisted flat band.

A mirrored scale on an analog meter permits the technician to reduce the problem of parallax.

Some basic troubleshooting tests for amplifiers and battery condition were given in this chapter. Also, ways to check calibration were included.

Ways of preventing meter loading were discussed. A quick way of determining meter resistance was described.

Some of the measuring techniques used for voltmeters are also useful with oscilloscopes.

PROGRAMMED SECTION

Instructions for using this programmed section are given in Chap. 1

---

1. Which of the following tests can be used to distinguish between a silicon diode and a germanium diode?

  A. The forward-to-reverse resistance ratio must be 10 to 1. Go to block 7.

  B. Measure the forward voltage across the diode on the DC voltage scale. Go to block 14.

---

2. You have selected the wrong answer for the question in block 27. Read the question again. Then go the block with the correct answer.

---

3. The correct answer to the question in block 16 is B. The voltage is about

$$\frac{5}{15} \times 30V$$

or one-third of 30V.

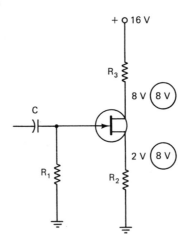

Here is your next question:

For the Class A amplifier circuit shown in this block, the numbers not circled are given by the manufacturer and the circled values are actual measurements. Which of the following is correct?

  A. The JFET is saturated. Replace it. Go to block 18.

  B. Replace R_1 because the JFET is saturated. Go to block 11.

---

4. You have selected the wrong answer for the question in block 26. Read the question again. Then go to the block with the correct answer.

---

5. The correct answer to the question in block 24 is B. Tapping the center tap of the volume control—with the control set to its maximum sound position—should produce clicking noises in the speaker if the audio section is OK.

Here is your next question:

The emitter-base short-circuit test should not be used with

A. amplifiers having a high gain. Go to block 28.

B. direct-coupled amplifiers. Go to block 23.

---

6. You have selected the wrong answer for the question in block 12. Read the question again. Then go to the block with the correct answer.

---

7. You have selected the wrong answer for the question in block 1. Read the question again. Then go to the block with the correct answer.

---

8. The correct answer to the question in block 14 is B. The sticking friction is the reason the meter should be lightly tapped when adjustments produce small changes in the output.

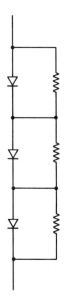

Here is your next question:

The resistors across the diodes in the illustration for this block are used to equalize the reverse voltages across the diodes. These resistors have a very high resistance value. Measure the resistance of each resistor using

A. the low-power ohms feature. Go to block 24.

B. the higher-power ohms scale. Go to block 17.

9. You have selected the wrong answer for the question in block 29. Read the question again. Then go to the block with the correct answer.

10. You have selected the wrong answer for the question in block 24. Read the question again. Then go to the block with the correct answer.

11. You have selected the wrong answer for the question in block 3. Read the question again. Then go to the block with the correct answer.

12. The correct answer to the question in block 23 is B. Measure the battery voltage when it is connected into its circuit. When the circuit is energized, the battery terminal should not drop to less than 80 percent of the full-charge value. (The 80 percent limit is typical, but you should use your own limit for units you are familiar with.)

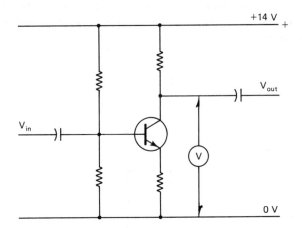

Here is your next question:
The voltmeter in the illustration for this block indicates the collector voltage is +12V. Therefore, this Class A amplifier is

A. nearly cut off. Go to block 26.

B. nearly saturated. Go to block 6.

C. working properly. Go to block 20.

13. You have selected the wrong answer for the question in block 26. Read the question again. Then go to the block with the correct answer.

14. The correct answer to the question in block 1 is B. Both germanium and silicon diodes should have a front-to-back ratio of 10 to 1, so the test described in A of question 1 will not tell the type of diode. Silicon diodes have a forward voltage drop of 0.7V. Germanium diodes have a forward drop of 0.2V.

Here is your next question:
Stiction is a problem with

A. digital multimeters. Go to block 15.

B. analog multimeters. Go to block 8.

---

15. You have selected the wrong answer for the question in block 14. Read the question again. Then go to the block with the correct answer.

---

16. The correct answer to the question in block 29 is B. The difference between -4 dB and $+2$ dB is 6 dB.

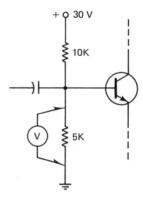

Here is your next question:
Estimate the voltage indicated by the voltmeter for the circuit in this block.

A. 5V. go to block 25.

B. 10V. Go to block 3.

---

17. You have selected the wrong answer for the question in block 8. Read the question again. Then go to the block with the correct answer.

---

18. The correct answer to the question in block 3 is A. There should be a voltage drop across the JFET, but R_2 is not a likely cause of the problem. Using the statistical chart, the most likely cause is the JFET.

Here is your next question:
For the circuit shown, assume there is no input signal. In this block you would expect the collector voltage to be about

A. 12V. Go to block 27.

B. 0V. Go to block 21.

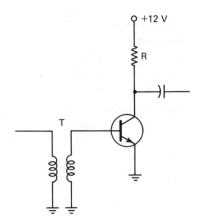

---

19. You have selected the wrong answer for the question in block 23. Read the question again. Then go to the block with the correct answer.

---

20. You have selected the wrong answer for the question in block 12. Read the question again. Then go to the block with the correct answer.

---

21. You have selected the wrong answer for the question in block 18. Read the question again. Then go to the block with the correct answer.

---

22. You have selected the wrong answer for the question in block 29. Read the question again. Then go to the block with the correct answer.

---

23. The correct answer to the question in block 5 is B. The emitter-base short-circuit test can destroy a direct-coupled amplifier.

Here is your next question:

Which of the following tests is used for checking a battery?

A. Use a digital voltmeter with a high input resistance. Go to block 19.

B. Measure the voltage while the battery is delivering its normal load current. Go to block 12.

---

24. The correct answer to the question in block 8 is A. The low-power ohms scale is used so that the ohmmeter will not turn the diodes on. Of course, the measurement could be made by reversing the ohmmeter leads, but a careless error is possible with that measurement.

Here is your next question:

A technician taps a volume control lead with a screwdriver. This test is to check the

A. AGC circuit. Go to block 10.

B. audio amplifiers. Go to block 5.

---

25. You have selected the wrong answer for the question in block 16. Read the question again. Then go to the block with the correct answer.

---

26. The correct answer to the question in block 12 is A. The amplifier current is too low, so there is not enough voltage drop across the collector resistor. Therefore, the voltage at the collector is too high. It should be about one-half the power supply voltage.

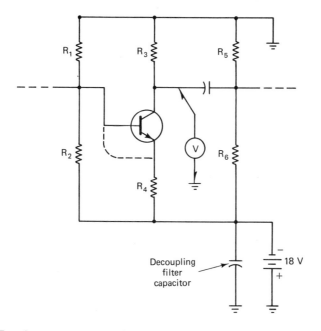

Here is your next question:

Consider the Class A amplifier shown in this block. An emitter-to-base short connected to the transistor will cause the voltmeter to display

A. −18V. Go to block 13.

B. +18V. go to block 4.

C. 0V. Go to block 29.

---

27. The correct answer to the question in block 18 is A. There is no forward bias on the transistor, so this is a Class B operation. The transistor is cut off when there is no signal, so there is no drop across R and the collector voltage will be the same as the supply voltage.

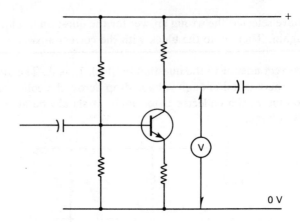

Here is your next question:
Refer to the circuit in this block.
With no input signal the voltmeter displays +6V. When a pure sine wave voltage applied to V_{in}, the voltmeter should display

A. a higher voltage. Go to block 2.

B. a lower voltage. Go to block 30.

C. no change. Go to block 31.

---

28. You have selected the wrong answer for the question in block 5. Read the question again. Then go to the block with the correct answer.

---

29. The correct answer to the question in block 26 is C. The emitter-to-base short turns the transistor off. With no current flowing through R_3 there is no voltage drop across that resistor. That puts the collector at common potential, or zero volts.

Here is your next question:
On the dB scale, a meter shows −4 dB at the output of a transmission line that terminated in a load resistance of 600 ohms. At the 600-ohm input the meter shows +2 dB. What is the dB loss in the line?

A. 4 dB. Go to block 22.

B. 6 dB. Go to block 16.

C. Neither choice is correct. Go to block 9.

---

30. You have selected the wrong answer for the question in block 27. Read the question again. Then go to the block with the correct answer.

---

31. The correct answer to the question in block 27 is C. This is the sine wave distortion test described in the chapter.

Here is your next question:

When a radio is tuned from no station to a station, the AGC (AVC) voltage should

A. become more positive. Go to block 32.

B. become more negative. Go to block 33.

C. More information is needed to answer. Go to block 34.

32. You have selected the wrong answer for the question in block 31. Read the question again. Then go to the block with the correct answer.

33. You have selected the wrong answer for the question in block 31. Read the question again. Then go to the block with the correct answer.

34. The correct answer to the question in block 31 is C. It is necessary to know if forward or reverse AGC (AVC) is being used.

You have now completed the programmed section.

TEST YOUR KNOWLEDGE

1. Which of the following is less susceptible to extraneous noise? ___
   A. digital multimeter
   B. analog multimeter
2. In order to avoid turning on a semiconductor junction when making a resistance measurement, use the ___ feature.
3. To measure AC current with a volt-ohm-milliammeter so that the voltage scale can be read directly as current, use a resistor having a resistance of ___.
4. What is the approximate input resistance of a VOM on the 100V scale if its meter movement is 50,000 ohms per volt? ___
5. Current-reading meters can be compared by connecting them ___.
   A. in series in a circuit.
   B. in parallel in a circuit.
6. Demodulator probes are used when the modulated signal being measured ___.
   A. is above the frequency range of the meter.
   B. does not have a 0V average value.
7. What scale should be avoided when connecting an ohmmeter across a PN junction? ___
8. Which of the following is more rugged? ___
   A. taut-band meter movement
   B. jeweled meter movement
9. Do not use the short-circuit test in the following cases: ___ and ___.
10. What type of signal is used for the voltmeter distortion test? ___

ANSWERS
TO TEST YOUR KNOWLEDGE

1. B
2. low-power ohms
3. 1 ohm
4. 5 megohms
5. A
6. A
7. R $\times$ 1
8. A
9. direct-coupled amplifiers and Class B amplifiers
10. sine wave

4

Servicing

with an Oscilloscope

CHAPTER OVERVIEW

An oscilloscope is the most versatile test instrument on the workbench. Here are some of the measurements and displays that an oscilloscope can perform:

MEASUREMENTS

Current
Voltage
Time intervals
Rise time and decay time
Bandwidth
Frequency and frequency differences
Percent modulation
Phase angles

DISPLAYS

Signals in either a time domain or a frequency domain
Waveform distortion
Linear distortion of amplifiers
Frequency distortion of amplifiers
Phase distortion of amplifiers

Frequency response of amplifiers

Characteristic curves of components

Alphanumeric characters

Timing of signals

Despite its usefulness in performing a wide variety of service tasks, some technicians prefer to use a voltmeter instead of an oscilloscope. One of the most important disadvantages of the oscilloscope is its physical size. It is not as portable (or at least as conveniently portable) as a voltmeter. So, for troubleshooting in the home, the voltmeter is still favored by some technicians.

Very small portable oscilloscopes are now available for the type of servicing that requires their use. One popular type fits in a briefcase, and one manufacturer is making a hand-held type. As a rule, these are very expensive when compared to the cost of a volt-ohm-milliammeter.

An oscilloscope is more difficult to use than a voltmeter. That is no obstacle for a good technician, but many busy technicians cannot take the time to explore its full range of applications. In this chapter some of the techniques for using scopes for servicing are discussed.

There is one mistake made with oscilloscopes that could be easily eliminated: Technicians don't read the manual that comes with the equipment—a mistake that occurs with all test equipment.

Objectives

Some of the topics discussed in this chapter are listed here:

- What is a frequency domain display and how is it used?
- What is a time domain display and how is it used?
- When is an alternate mode used with oscilloscopes?
- When is a chop mode used with an oscilloscope?
- Why is a delayed sweep necessary?
- How can a half-cycle average be determined from a waveform display?
- How can an RMS value be determined from a waveform display?

COMPARISON OF OSCILLOSCOPE TYPES

Recurrent sweep scopes are by far the cheapest and easiest to use.

A block diagram of a typical recurrent sweep oscilloscope is shown in Fig. 4-1. In this type, a sawtooth waveform is used to produce a linear horizontal sweep. That sweep is synchronized with the incoming signal by a control called *sync amplitude*. It samples the incoming signal on the vertical input terminals and uses it as the synchronizing signal.

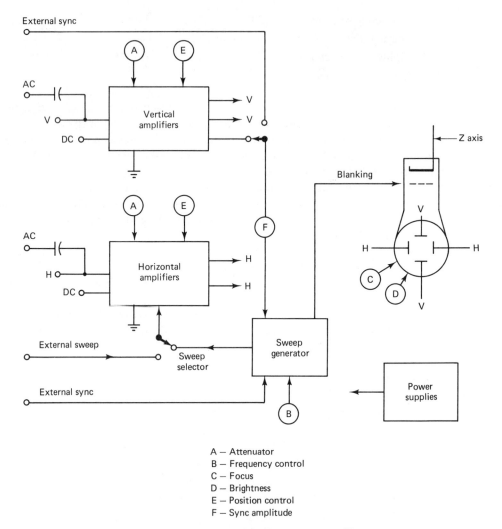

External sync

AC

V

DC

Vertical
amplifiers

V
V

AC

H

DC

Horizontal
amplifiers

H
H

External sweep

Sweep
selector

External sync

Sweep
generator

Blanking

Z axis

Power
supplies

A — Attenuator
B — Frequency control
C — Focus
D — Brightness
E — Position control
F — Sync amplitude

Figure 4-1 Block diagram of a recurrent sweep oscilloscope.

In the normal use of an oscilloscope, the sync control must be adjusted for the *minimum* amount of sync signal necessary to lock the horizontal oscillator onto the waveform being observed. Excessive sync input will cause distortion of the waveform being displayed.

As shown in the block diagram, it is possible to use an external signal to synchronize the internal sweep. It is also possible to deliver an external sweep signal with the internal sawtooth generator disconnected.

Both the vertical and horizontal amplifiers have two choices of inputs: DC and AC. With the AC input, the signal goes through a capacitor. That prevents any DC offset voltage from getting into the amplifier. In some cases, that DC offset could be sufficient to deflect the beam off the face of the scope.

In the scope of Fig. 4-1, the sweep generator output passes through the horizontal amplifiers. That way the attenuator and position controls of that amplifier can be used to control the sweep.

The blanking control shuts the beam off during retrace.

Dual-trace scopes should not be confused with *dual-beam* scopes, which have a separate electron beam for each trace.

Memory scopes can display a waveform and retain that waveform over a period of time. Earlier memory scopes employed a special cathode ray tube to retain the display.

Modern memory scopes break the waveform into numbered bits. This technique is the same as is used in compact discs. It is called pulse code modulation (PCM). The numbered parts that represent the waveform are stored in a memory and can be recalled at any time.

Digital logic scopes are specifically designed to show logic timing diagrams. They are characterized by having a large number of traces, usually six or more.

In addition to the scopes just mentioned, some specialized oscilloscopes are

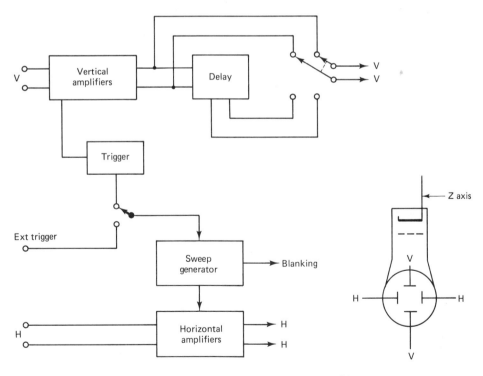

Figure 4-2   Block diagram of a triggered sweep oscilloscope.

readily available. As an example, the phase relationships between signals in a color television receiver are very important for proper display of the color picture. Special oscilloscopes—called *vector scopes*—display and measure these phase relationships.

Radar displays are also specialized oscilloscopes used for measuring distance, altitude, and angles.

Triggered sweep scopes require an input signal to start the trace. A block diagram of a typical triggered sweep scope is shown in Fig. 4-2. Since the signal starts the trace, the display is very stable. In a typical triggered sweep scope, the time base is marked in seconds, milliseconds, or microseconds. In the calibrate function, these periods of time are very accurately controlled.

Except for the triggering feature, the block diagrams of the recurrent and triggered sweep scopes are very similar.

The control for the horizontal sweep is marked in units of time. Accurately controlled time bases make it easy to measure the frequency. The procedure is to measure the time (T) for one cycle, then calculate the frequency using the equation

$$f = \frac{1}{T}$$

Dual-trace scopes can display two time bases at the same time. In most common oscilloscopes this is accomplished by using an internal electronic switch that switches back and forth between the two waveforms being displayed. The switching is so rapid that the eye perceives the traces as being on at the same time.

Oscilloscope Bandwidth

Oscilloscopes are advertised and sold according to the bandwidth of frequencies that can pass through their vertical amplifier. Remember that bandwidth is the range of frequencies between points where the amplitude drops to 70.7 percent of its maximum voltage value. This is shown in Fig. 4-3.

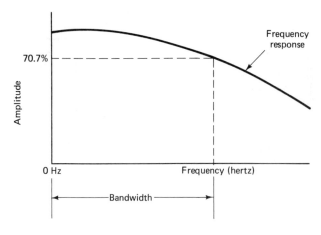

Figure 4-3 Bandwidth is the range of frequencies between points where the voltage drops to 70.7 percent of maximum. In this illustration, bandwidth is measured from 0 hertz.

When the characteristic curve shows power vs. frequency, the bandwidth is the range of frequencies between the points where the curve drops to 50 percent of maximum.

Bandwidth is a very important consideration, especially when observing non-sinusoidal inputs. Consider the case of a square wave signal. As shown in Fig. 4-4, a square wave consists of a fundamental pure sine wave form and a large number of odd

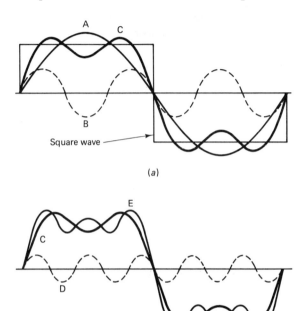

(a)

(b)

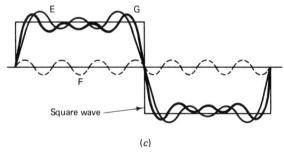

(c)

A — Fundamental
B — Third harmonic
C — Fundamental plus third harmonic
D — Fifth harmonic
E — Fundamental plus third and fifth harmonics
F — Seventh harmonic
G — Fundamental plus third, fifth, and seventh harmonics

Figure 4-4 Makeup of a typical square wave. In reality, the harmonics would extend to infinity to make a perfect square wave.

sine wave harmonics. In order to reproduce a reasonable display of the waveform, it is necessary for the vertical amplifier of the oscilloscope to pass at least eight or ten of the harmonic frequencies. If an amplifier cannot pass the harmonic frequencies, it will distort the square wave form. This is the basis for the qualitative square wave test described in Chap. 1.

Another reason for needing a wide bandwidth is the reproduction of a glitch. A glitch is an undesired transient voltage having a very short time period. For example, a glitch may have a period of only 10 or 15 nanoseconds. In order to display this glitch, the vertical amplifier must be able to show the very short pulse. That, in turn, means it must pass high harmonic frequencies.

Two limiting factors must be taken into consideration. One is the persistence of the scope screen. Persistence is a measure of how long a trace is visible after the electron beam passes. If the persistence is too long, you won't be able to see changes in the waveform being displayed. The second important factor is the bandwidth of the vertical input amplifier. It must pass all of the harmonics that produce that pulse wave form. If the vertical amplifier cannot pass the range of harmonics associated with the glitch, you will not see it displayed.

A simple circuit that you can construct to determine your scope's ability to display a glitch is discussed in Chap. 10.

Delay

The need for a delay in the vertical amplifier circuit shown in the block diagram of Fig. 4-2 is very apparent when you are looking at a square wave. As shown in Fig. 4-5, you can't see the rise time of the first cycle when the scope doesn't have a built-in delay.

You could decrease the sweep speed so you can see two cycles. That way you can see the rise time of the first cycle. However, the rise time is easier to measure—because it takes up a greater part of the trace—with the single pulse display. This is especially true if you are using the oscilloscope sweep expander.

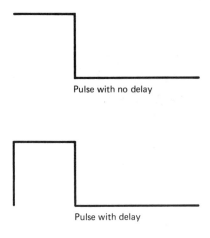

Pulse with no delay

Pulse with delay

Figure 4-5 Unless there is a time delay circuit in the amplifier, you will not be able to see the leading edge of a step function.

FREQUENCY VS. TIME DOMAIN

Waveforms have three dimensions: amplitude, frequency, and time elapsed. Normally, it is only possible to graph two of these dimensions on a flat piece of paper or graph, so the two most commonly used graphs are given.

Figure 4-6 shows the relationship between the two-dimensional views as they are seen on an oscilloscope screen. The three-dimensional view in Fig. 4-6(a) has only two representative sine wave forms, but any waveform would be located on the graph.

As shown in Fig. 4-6(b), the time axis view has the later times as you look from left to right. This is the time domain display most often seen on scopes.

The *frequency domain display* shown in Fig. 4-6(c) is also available on an oscilloscope. Uses of frequency domain displays are discussed in Chap. 5.

SCOPE ACCESSORIES

In addition to the scopes for displaying waveforms, there are usually some provisions for measuring frequency, time, and voltage.

Be sure to study the methods of measuring voltage and frequency described in

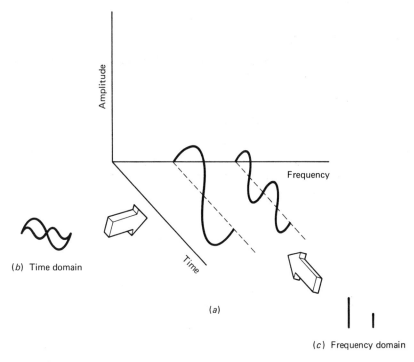

Figure 4-6   This illustration shows how time delay and frequency delay displays are made.

the manual for your triggered sweep oscilloscope. Those internally calibrated features should be checked periodically to ensure that your equipment is working properly.

Voltage Calibration

To calibrate an oscilloscope, or to verify its internal calibration, use the same techniques described for voltmeters in Chap. 3. Use the volts-per-division calibration.

A 1.5V battery can be used to calibrate the DC vertical input (DC) terminal. A filament transformer secondary can be used to check the AC calibration.

A variable AC transformer—called a variac—is also very useful for checking AC calibration.

On many oscilloscopes there is a calibration voltage output terminal on the front panel of the instrument.

Some technicians prefer to calibrate their instruments with a focused dot, rather than a trace, on the screen. Keep the intensity of that dot as low as possible for best focus.

CAUTION

If you leave a bright dot on the CRT screen you can get a permanent burn mark in the phosphor coating. *Never* leave the dot on the screen for a long period of time.

Sample Problem

The variable transformer in Fig. 4-7 is adjusted for a voltmeter reading of 10V AC. What peak-to-peak value should be indicated on the oscilloscope?

Solution

The peak-to-peak reading should be

$$2 \times 1.414 \times 10 = 28.28V \text{ AC}$$

Remember that the voltmeter indicates the RMS value, or 1.414 times the peak value! Note the setup for checking voltage calibration shown in Fig. 4-7.

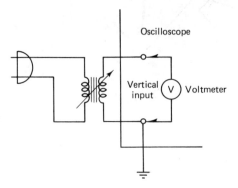

Figure 4-7   Illustration for sample problem.

Frequency Calibration

If a scope is set for 10 milliseconds per horizontal division, there should be about 1.6 divisions for one cycle of 60 hertz AC voltage.

To get an accurate number of milliseconds per division, a frequency counter can be used, as shown in Fig. 4-8. Once the time base calibration has been checked on one of the divisions, you can assume that the other divisions on the scope time base are also accurate.

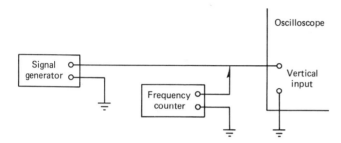

Figure 4-8 Checking the oscilloscope time base for accuracy.

Chop and Alternate

Dual-trace triggered sweep oscilloscopes have a built-in electronic switch. Its purpose is to switch back and forth between the two displays so rapidly that your eye perceives them to be on at the same time. Electronic switches can also be used with recurrent sweep scopes as an external accessory.

In reality, the electronic switch is nothing more than a square wave generator that switches the trace back and forth between two DC levels.

Regardless of whether the electronic switch is built-in or used externally, there are usually two modes of operation: chop and alternate.

In the *alternate* mode, one line is traced and then the other line is traced. Each trace is displayed alternately.

At very low frequencies this alternate switching produces a distracting pulsed display. In that case, the *chop* mode should be used. In the chop mode a portion of one waveform is shown on one trace and then a portion of the second waveform is shown on the other trace. The back-and-forth switching is accomplished very rapidly so that each trace is broken into small parts. This is illustrated in Fig. 4-9.

Low-Capacity Probes

For most applications, technicians use a low-capacity probe. This puts a capacitor in series with the probe, as shown in Fig. 4-10.

Remember that two series capacitors have a combined capacity that is lower than either capacity value. Therefore, the series capacitor combines with the scope capacity to *reduce* the total capacity as seen by the input signal.

In the chop mode the scope switches back and forth rapidly between the two low frequencies.

In the alternate mode the scope displays one complete line and then the other complete line.

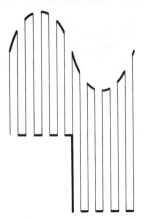
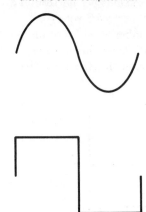

Figure 4-9 Comparison of chop and alternate displays. In the chop mode the trace moves back and forth quickly between parts of the waveform. In the alternate mode one complete waveform is traced and then the other one is traced.

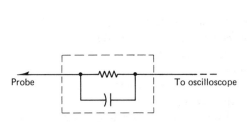

Figure 4-10 Circuit for a low-capacity probe.

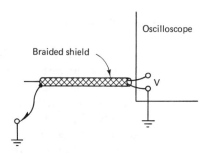

Figure 4-11 Shielded wire should never be grounded at both ends.

For low-frequency applications, such as the low-frequency end of the audio spectrum, a direct probe can be used.

A common mistake is to ground the shield of a scope lead coaxial cable at both ends. Figure 4-11 illustrates the problem. A very small difference in the common voltage between the scope and the system being measured will cause an AC current to flow in the shield. That, in turn, causes an electromagnetic field that injects an undesired interference signal in the scope.

Parallax in Scope Measurements

If you are using an add-on gradicule for either frequency or voltage measurements, you must be careful not to introduce a parallax error. (See the discussion of parallax in Chap. 3)

To avoid the problem of parallax, some special-application oscilloscopes have the gradicule marked on the inside of the CRT face. Ordinary oscilloscopes do not have this feature, so it is up to you to avoid the error.

Measuring Current

The procedure for measuring AC current on an oscilloscope is the same as that for measuring voltage. A 1-ohm resistor is placed in series with the current to be measured. The voltage across that resistor is measured with a scope, and the current is numerically equal to the voltage.

If the current being measured is too small to get a sufficient voltage across 1 ohm, 10 ohms or 100 ohms can be used, provided the circuit resistance is at least ten times greater than the series resistance inserted.

In order to get the maximum accuracy with scope measurement of a DC or pulse waveform, you should adjust the scope vertical positioning control so that the trace line is at the bottom of the gradicule. This is shown in Fig. 4-12. That gives you the full height of the gradicule for making the measurements. It is not a good idea to make measurements using a dot (rather than a trace) on the scope.

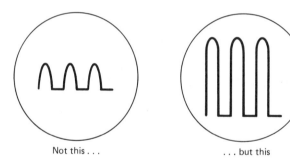

Not this . . .   . . . but this

Figure 4-12 Get the maximum accuracy of scope and measurements by making use of the full height available.

Determining Average and RMS Value

You will remember that voltmeters are usually calibrated to read RMS values of sine wave voltages and currents. Some types of meters are designed to measure true RMS values, for example, the thermocouple meter.

You can measure the average and RMS values of a nonsinusoidal wave with a scope if you are willing to do a few simple calculations. Figure 4-13 shows how the average value is obtained, and Fig. 4-14 shows how the RMS value is obtained. It is only necessary to take between five and ten measurements to get a fairly accurate determination of the average or RMS value. Of course, the greater number of measurements the more accurate the results.

The average value is calculated the same way as finding an average of anything else. Assume the waveform in Fig. 4-13 is for a voltage. Add the voltages, as shown, for

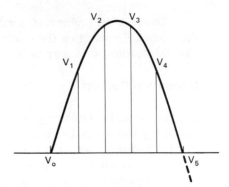

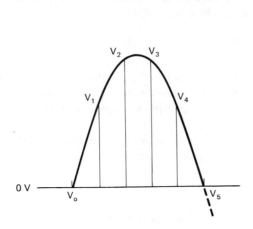

Average value = $\dfrac{V_1 + V_2 + V_3 + V_4 + V_5}{5}$.

Note: Do not include V_o.

Figure 4-13 Average value is taken by averaging individual voltage values.

1. Square each value of V.
2. Add the squared values.
3. Take the average of the squared values.
4. Take the square root of that average.

RMS value = $\sqrt{\dfrac{V_1^2 + V_2^2 + V_3^2 + V_4^2 + V_5^2}{5}}$.

Note: Do not include V_o.

Figure 4-14 The RMS value is obtained by the method shown here.

one cycle. Then divide by the total number of voltages that you added to get the average value.

The procedure for finding the RMS value is to work backward from the term root mean square:

1. *Square* each value.
2. Find the average, or *mean*.
3. Take the *square root* of the result.

As with the average value, more values—that is, divisions—means more accuracy. But you may not need that amount of accuracy for what you are doing.

In most equipment the manufacturer gives peak-to-peak values of voltages (or currents) when the RMS value is needed. However, if you are working in a situation where you need to know the average or RMS value, the technique shown in Fig. 4-13 and 4-14 works reasonably well.

MEASURING OR EVALUATING COMPONENTS WITH AN OSCILLOSCOPE

Oscilloscopes are ideal for checking the PN junction of diodes, transistors, and other semiconductor devices. They can also be used to determine the operating characteristics of thyristors and other components. The general procedure is to display the current through the device on one axis and the voltage across the device on the other axis. (The method of displaying current has been discussed earlier in this chapter.) Some oscilloscopes have a *component test* feature. If your scope doesn't have one, it is a simple matter to make some basic test equipment for doing the job. This test equipment is described in Chap. 11.

AMPLIFIER SERVICING WITH AN OSCILLOSCOPE

There are two types of distortion in amplifiers: frequency distortion and linear distortion. Phase distortion is included in the category of frequency distortion because the amount of phase distortion is dependent on the frequency of operation.

Frequency Distortion

Frequency distortion occurs when an amplifier can't pass the range of frequencies it is designed for. The square wave test and the sawtooth test are qualitative. They permit a quick overview of the amplifier's ability to pass its range of frequencies.

The point-by-point test shown in Fig. 4-15 is accurate but time consuming. The signal generator is tuned over the range of frequencies. The amplitude of the output is plotted for each frequency.

The resulting graph is often plotted on semilog paper. It is linear in the vertical direction and logarithmic in the horizontal direction. This permits details over a wide range of frequencies to be explored.

For each measurement taken, it is necessary to ensure that the input amplitude

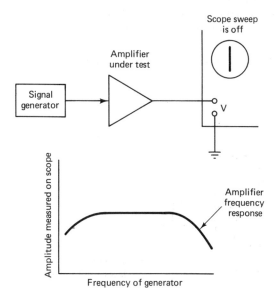

Figure 4-15 The amplifier frequency response can be obtained by making point-by-point measurements and then graphing the result. Notice that only a vertical line is used for making the voltage measurement.

has not changed when the frequency was changed. This step is not necessary for the more expensive signal generators that have regulated output amplitudes.

For a quick test the graph need not be plotted. Just run the generator frequency control over the range of required frequencies while observing the amplitude displayed on the scope. Turn off the horizontal amplitude control for this test since you are only interested in amplitude.

(Remember the caution against leaving a dot on the screen of the scope.)

With the generator connected to the amplifier, observe the input to the amplifier on the scope. The length of the line (representing amplitude) should be constant throughout the range.

Then connect the scope to the output and adjust the generator through the range again. The bandwidth is the range of frequencies where the length of the line does not drop below 70 percent of maximum.

If you see a bright spot on either end—or both ends—of the line, it means the output wave is distorted and the test is not valid. Reduce the input signal amplitude and try again.

The frequency response curve of an amplifier can be plotted directly on an oscilloscope by using the scope in the frequency domain. This procedure is described in Chap. 5.

Linear Distortion

Linear distortion occurs in an amplifier when the output waveform is not an exact representation of the input waveform. The input and output waveforms can be out of phase by 180 degrees, but they must have exactly the same waveform in a low-frequency linear amplifier. Class A amplifiers should have no linear distortion.

High-frequency linear amplifiers of the type used in transmitters may not be operated Class A. They are designed to produce a very high gain. So a "linear" amplifier in a transmitter system may actually be a Class B amplifier or, in some cases, even a Class C operation.

The principle of operation is shown in Fig. 4-16. The input signal to this Class B

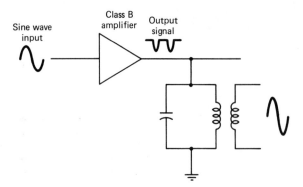

Figure 4-16   This Class B amplifier produces a full sine wave output.

amplifier is a pure sine wave. The output is a half-wave rectified sine wave. It does not look anything like the sine wave input. This distortion is typical of Class B amplifiers.

The tank circuit in the output of the Class B amplifier is resonant to the same frequency as the input signal. The half-wave variations produce high-amplitude circulating currents in the tank. That signal is coupled to the transformer secondary as a pure sine wave.

Assume that the amplifier is linear and operating in the low-frequency range. One quick test for linearity uses a lissajous pattern. It is a good idea to review the generation of lissajous patterns before describing this test.

Figure 4-17 shows a test setup to produce a lissajous pattern on an oscilloscope. The scope is set so that the horizontal input signal replaces the internal sweep circuitry. (In other words, there is no sawtooth time base on the scope.)

The horizontal deflection of the beam is produced by the signal delivered to the horizontal input terminals. On some oscilloscopes this is accomplished by turning the internal sweep off, or to *external.* In other scopes this is called the X-Y operation.

In order to get a true lissajous pattern, it is necessary that the signal at the horizontal plates produce exactly the same deflection as the signal at the vertical plates. This is accomplished by feeding the two signals to the scope, one at a time, and adjusting for a specific amount of deflection.

NOTE

Lissajous tests are *not* valid unless the two inputs are adjusted—one at a time—so that they produce the same amount of deflection on the oscilloscope.

The test setup in Fig. 4-17 should produce a perfect circle if the two frequencies are identical and they are 90 degrees out of phase. This provides a test to determine if there is distortion in a sine wave. If you are sure that one of the two inputs is a perfect sine wave, then you can determine if the other is a perfect sine wave (or a distorted sine wave) by using this test. *If you can't get a perfect circle, the unknown waveform is not a pure sine wave.*

If the phase angle is other than 90 degrees, the lissajous pattern will be somewhere between an ellipse and a straight line. Some of the variations are shown in Fig. 4-18.

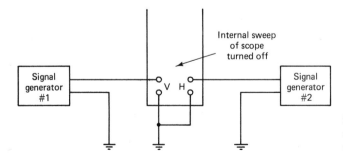

Figure 4-17   The lissajous patterns on an oscilloscope can be obtained using this test setup.

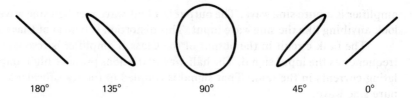

Figure 4-18 Various lissajous patterns obtained when the waveforms are pure sine wave and equal in frequency. Only the phase between the two waves is different.

Because the out-of-phase sine wave produces a specific pattern, it is possible to determine the phase difference between the two sine waves as a quantitative measurement. The procedure is shown in Fig. 4-19.

With the two waves producing the ellipse, center the ellipse on the X-Y axis. To do this, count from the center in each direction horizontally and from the center in each direction vertically to make sure that the X and Y axes are crossing in the center. Then measure A and the projection of B. Use those values in the equation given in the illustration. If you have a scientific calculator, simply divide A by B. Then push the inverse sign button and the phase angle will be displayed.

You can use the lissajous pattern to calibrate a signal generator frequency against the frequency of one that is known to be accurate.

If the input and output frequencies are not the same, the lissajous pattern takes on more complicated displays, as shown in Fig. 4-20. You can find the ratio of the vertical to the horizontal frequency using the simple technique shown in Fig. 4-21.

Determining frequency ratios by the method shown in Fig. 4-21 is an interesting trick that you are expected to know. However, it has no practical application in troubleshooting.

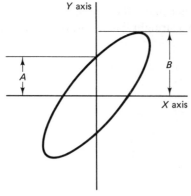

$$\text{Phase angle} = \phi = \sin^{-1}\frac{A}{B}, \quad \text{or} = \arcsin\frac{A}{B}.$$

Figure 4-19 The actual phase difference between two waveforms can be obtained from the lissajous pattern. The procedure is shown here.

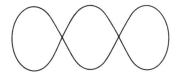

Figure 4-20 A lissajous pattern formed by a 3-to-1 frequency ratio.

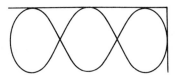

1. Draw horizontal and vertical lines as shown.

2. $\dfrac{\text{No. of times pattern touches horizontal line}}{\text{No. of times pattern touches vertical line}} = \dfrac{\text{Vertical frequency}}{\text{Horizontal frequency}}$

3. $\dfrac{\text{Vertical frequency}}{\text{Horizontal frequency}} = \dfrac{3}{1}$

The vertical frequency is three times the horizontal frequency.

Figure 4-21 The procedure for finding the ratio is shown here.

The Amplifier Lissajous Test

Having reviewed the principle of the lissajous pattern, we can now use the pattern to test for amplifier distortion. The setup is shown in Fig. 4-22.

Note that the weaker input signal of the amplifier goes to the oscilloscope vertical input. In many scopes the vertical input has more gain, so it is better able to amplify that weaker signal.

Remember that the horizontal and vertical deflections must be equal to get a

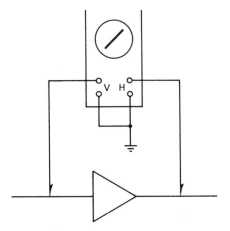

Set vertical and horizontal deflections equal. Then set scope horizontal to external input (may be called X-Y input).

Figure 4-22 This lissajous setup can be used for finding distortion in an amplifier.

lissajous pattern. *The pattern should be a straight line when the frequency used is about 1000 Hz.*

At higher frequencies the pattern will open to an ellipse showing some phase shift distortion. The specifications for the amplifiers will determine how much phase shift distortion can be tolerated and at what frequency.

The human ear is not sensitive to phase shift distortion, so this does not usually represent a serious problem with audio amplifiers. For other amplifiers it may be very important.

As shown in Fig. 4-23, nonlinear distortion will produce a pattern other than a straight line. Remember that bright ends on either or both ends of the pattern indicate distortion.

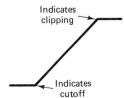

Indicates clipping

Indicates cutoff

Figure 4-23 Nonlinear distortion gives a lissajous pattern that is not a straight line.

This test works best with a single frequency input. If you are looking at a range of frequencies, such as the output signal from an audiotape, then the various frequencies will produce various amounts of phase distortion. You won't get a straight line. That can still be, however, a very valuable test.

Intermodulation Distortion

Intermodulation distortion is another form of nonlinear distortion. It occurs when two different frequencies are applied to an amplifier and those frequencies heterodyne. Heterodyning or modulation can only take place in a nonlinear stage. Therefore, if it does take place in an amplifier that is supposed to be linear, nonlinear distortion is occurring.

Typically, an intermodulation distortion test involves putting two frequencies into the amplifier and then looking at the output on a frequency domain display. See Fig. 4-24. A special piece of test equipment—called a *spectrum analyzer*—can be used to simplify this test.

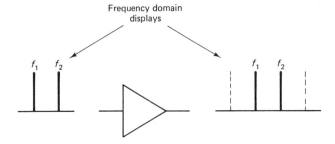

Frequency domain displays

f_1 f_2

f_1 f_2

Figure 4-24 A simplified illustration showing the effect of intermodulation distortion.

Intermodulation distortion is very undesirable because it produces frequencies that are not in the original signal. However, there is some controversy about this test. The question is: Which two frequencies do you use for testing? Different technical groups support different frequencies. So, when you are looking at the specifications for an amplifier, the intermodulation distortion has to be defined in terms of the frequencies.

You might wonder why it is necessary to use tests like the lissajous and intermodulation. Why not just simply apply a sine wave to the input of an amplifier and then see if you get a sine wave at the output? The reason it can't be done is very important. The human eye cannot see distortion in a sine wave as readily as it can see distortion in a circle and on intermodulation patterns. As a matter of fact, the sine wave can be distorted as much as 15 percent and still look quite acceptable to that human eye.

If you use the lissajous pattern distortion test on a function generator, you may be quite surprised at the amount of distortion present. The reason is that many function generators produce a "sine wave" by a frequency synthesis method. In this method the sine wave is actually constructed, point by point, *but it is not a pure sine wave.*

SUMMARY

Oscilloscopes are somewhat harder to use than voltmeters, but they can be used for a wider range of tests. Qualitative tests require you to make a judgment about the scope pattern you are observing. Quantitative tests provide you with exact values. Both qualitative and quantitative tests are obtainable with oscilloscopes.

Scopes are sometimes used to determine the bandwidth of an amplifier. You can get an approximate bandwidth by measuring the rise time of the square wave. A point-by-point plot is more useful and more accurate, but it takes longer.

Lissajous patterns can be used to determine if there is a pure sine wave. It can also be used to determine frequency ratios, but that technique has very little practical use.

The lissajous test can be used to determine if an amplifier is introducing linear distortion.

Harmonic distortion is more difficult to test in amplifiers. The test works better with a spectrum analyzer. The frequencies used for this test have not been standardized. Harmonic distortion is very important, but you must exercise great care in performing the test.

PROGRAMMED SECTION

The instructions for this programmed section are given in Chap. 1.

---

1. Shown in this block is a pattern obtained by an amplifier lissajous test for distortion. This pattern indicates that the amplifier

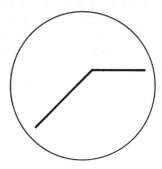

    A. is working properly. Go to block 7.

    B. has distortion. Go to block 14.

---

2. The correct answer to the question in block 14 is B. If you looked at the signal on a time base, you would see that the top and bottom of the wave are clipped. For severe clipping, it would look like a square wave.

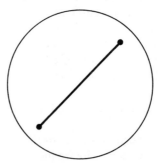

    Here is your next question:

To measure an AC current

    A. connect a 1-ohm resistor in series with the resistor. Measure the voltage across that resistor. The voltage and current are numerically equal. Go to block 15.

    B. measure the open-circuit voltage across a break in the line and divide by 10 megohms. Go to block 3.

---

3. You have selected the wrong answer for the question in block 2. Read the question again. Then go to the block with the correct answer.

4. You have selected the wrong answer for the question in block 23. Read the question again. Then go to the block with the correct answer.

5. The correct answer to the question in block 10 is A. The use of frequency domain displays is discussed in Chap. 5.

Here is your next question:

When looking at two very-low-frequency waveforms on a dual-trace scope, use the

A. alternate mode. Go to block 8.

B. chop mode. Go to block 23.

6. You have selected the wrong answer for the question in block 22. Read the question again. Then go to the block with the correct answer.

7. You have selected the wrong answer for the question in block 1. Read the question again. Then go to the block with the correct answer.

8. You have selected the wrong answer for the question in block 5. Read the question again. Then go to the block with the correct answer.

9. You have selected the wrong answer for the question in block 14. Read the question again. Then go to the block with the correct answer.

10. The correct answer to the question in block 15 is B. Your knowledge of the graphical method for finding RMS values makes it possible to understand this answer.

Here is your next question:

Two types of oscilloscope displays are time domain and

A. frequency domain. Go to block 5.

B. amplitude domain. Go to block 21.

11. You have selected the wrong answer for the question in block 22. Read the question again. Then go to the block with the correct answer.

12. You have selected the wrong answer for the question in block 13. Read the question again. Then go to the block with the correct answer.

13. The correct answer to the question in block 16 is B. The average value of a *half* cycle is 0.636 times the peak value.

Here is your next question:

Which of the following components is connected to the AC input terminal of an oscilloscope?

A. Capacitor. Go to block 22.

B. Resistive divider. Go to block 12.

14. The correct answer to the question in block 1 is B. The pattern indicates that the amplifier is cut off for a portion of the cycle.

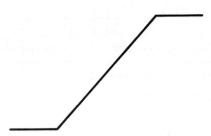

Here is your next question:

The lissajous pattern shown in this block was obtained for a certain amplifier. The pattern indicates that

A. the amplifier is working properly. Go to block 9.

B. the amplifier is clipping the sine wave. Go to block 2.

---

15. The correct answer to the question in block 2 is A. This is the same procedure as used for measuring current with a VOM.

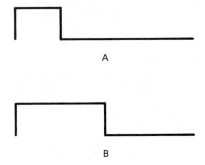

Here is your next question:

Refer to the waveforms in this block. Which has the higher RMS value?

A. The one marked A. Go to block 20.

B. The one marked B. Go to block 10.

C. Both have the same RMS value. Go to block 17.

---

16. The correct answer to the question in block 23 is A. A glitch contains a large number of harmonic frequencies. The vertical amplifier has to pass the harmonics in order to reproduce the glitch.

Here is your next question:

When there is no DC offset, the full-cycle average of a pure sine wave voltage is

A. 0.636 times the maximum voltage. Go to block 19.

B. zero volts. Go to block 13.

---

17. You have selected the wrong answer for the question in block 15. Read the question again. Then go to the block with the correct answer.

18. The correct answer to the question in block 22 is C. The built-in delay for this purpose should not be confused with the variable delay that is also available with some oscilloscopes.

Here is your next question:

What is a blanking circuit used for in an oscilloscope?

A. To remove portions of the trace. Go to block 28.

B. To turn off the beam during retrace. Go to block 25.

19. You have selected the wrong answer for the question in block 16. Read the question again. Then go to the block with the correct answer.

20. You have selected the wrong answer for the question in block 15. Read the question again. Then go to the block with the correct answer.

21. You have selected the wrong answer for the question in block 10. Read the question again. Then go to the block with the correct answer.

22. The correct answer to the question in block 13 is A. The capacitor prevents DC voltages and DC offsets from displacing the scope trace.

Here is your next question:

In order to make it possible to view the leading edge of the first cycle of square wave, some scopes use a

A. very wide sweep. Go to block 6.

B. sweep expander. Go to block 11.

C. delay circuit. Go to block 18.

23. The correct answer to the question in block 5 is B. In the chop mode the scope switches rapidly back and forth, showing part of each waveform with each alternation.

Here is your next question:

Which of the following is necessary for displaying a glitch on an oscilloscope?

A. Wide bandwidth at the vertical amplifier. Go to block 16.

B. High gain at the vertical amplifier. Go to block 4.

24. The correct answer to the question in block 25 is A. A VOM displays the RMS voltage. The RMS value is equal to $0.707 \times V_{max}$, or seven-tenths of the peak value—*not* the peak-to-peak value.

Here is your next question.

The horizontal trace of an oscilloscope is calibrated for 5 milliseconds per division. How many divisions are needed to display one cycle of a 100-H sine wave?

A. 2. Go to block 26.

B. 5. Go to block 29.

25. The correct answer to the question in block 18 is B. The purpose of blanking was discussed in the block diagram review of the oscilloscope.

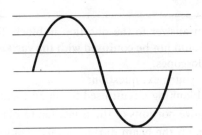

Here is your next question:
Refer to the oscilloscope display of a pure sine wave voltage. Assume the vertical attenuator is set for 5V per division. A VOM would measure this voltage to be

A. 10.6V. Go to block 24.

B. 21.2V. Go to block 27.

---

26. The correct answer in block 24 is A. Start by calculating the times (*T*) for one cycle:

$$T = \frac{1}{f} = \frac{1}{100} = 0.010 \text{ s}$$

Then convert to milliseconds:

$$0.010 \text{ s} \times \frac{1000 \text{ ms}}{1 \text{ s}} = 10 \text{ ms}$$

If one division takes 5 ms, then it will take two divisions for the complete (10 ms) display.

Here is your next question:
Bandwidth is the range of frequencies where the signal voltage drops to what percentage of maximum? Go to block 30.

---

27. You have selected the wrong answer for the question in block 25. Read the question again. Then go to the block with the correct answer.

---

28. You have selected the wrong answer for the question in block 18. Read the question again. Then go to the block with the correct answer.

---

29. You have selected the wrong answer for the question in block 24. Read the question again. Then go to the block with the correct answer.

---

30. 70.7 percent for voltage displays and 50 percent for power displays.

You have now completed the programmed section.

TEST YOUR KNOWLEDGE

1. What is another name for the Z axis of an oscilloscope? ____

2. The bandwidth of a circuit is the range of values where the *voltage* drops to ____ of maximum, or where the power drops to ____ of maximum.

3. Assume you are observing a pulse on an oscilloscope screen. Which part of the pulse might be lost if the sweep circuit of the scope does not have a delay feature? ____

4. What are the two most popular domains for oscilloscope delays? ____ ____

5. An AC voltmeter across the vertical input terminals indicates that the sine wave voltage is 20V. What peak-to-peak voltage is indicated by the scope? ____

6. What test equipment is useful for checking the calibration of an oscilloscope time base generator? ____

7. To view two low-frequency waveforms on a dual trace scope, use the ____.
   A. chop mode.
   B. alternate mode.

8. What components are used in a low-capacity probe? ____

9. Is the following statement correct? Always ground the braided shield of a shielded probe wire at both ends. ____

10. Which scope pattern is useful for checking an amplifier to determine if nonlinear distortion is present? ____

ANSWERS
TO TEST YOUR KNOWLEDGE

1. intensity modulation
2. 70.7 percent, 50 percent. Note: Your answers may have been *seven-tenths* and *one-half.*
3. the leading edge (rise time) of the pulse
4. time, frequency
5. 56+ volts

   Peak-to-peak voltage = RMS voltage $\times$ 1.414 $\times$ 2
   = 20 $\times$ 1.414 $\times$ 2

6. frequency counter
7. A
8. resistor and capacitor
9. not correct. Ground at the scope end only.
10. lissajous pattern

5

Signal Injection and Signal Tracing

CHAPTER OVERVIEW

The methods of signal injection and signal tracing are especially useful for trouble-shooting systems that have no output. A dead receiver is the example used in this chapter. However, the methods can be used in all electronic systems.

In addition to isolating trouble, it is sometimes necessary to evaluate the system using various tests and measurements. For example, the system may be able to pass a signal from input to output but cannot pass the required range of frequencies. Those necessary tests and measurements are also discussed in this chapter.

Objectives

Some of the topics discussed in this chapter are listed here:

- What are some methods of troubleshooting systems?
- What types of test equipment are useful for system troubleshooting?
- How is a frequency domain display used for system troubleshooting?
- What are some preliminary troubleshooting tests that can be done without test equipment?
- What are the precautions to be taken with regard to AFC and AGC/AVC?

SYSTEM TROUBLESHOOTING

This is opinion: *The best troubleshooting aid is the technician's knowledge of how a system (or circuit, or component) works.* If you know how something works, you can better understand what is happening when it isn't working.

Technicians sometimes take issue with this viewpoint. Their argument is that you cannot make any money in troubleshooting by analyzing a system. Some technicians even claim that students just out of school can't troubleshoot because they are too heavy in theory and too light in practical methods of finding (and repairing) troubles.

Perhaps the real truth about troubleshooting is somewhere between the two extremes: Quicker techniques can be used for troubleshooting in most cases, but for those few tough dogs that do not yield to fast troubleshooting methods, the best attack is an analysis based on the technician's knowledge of the system.

In this book a system is considered to be a combination of amplifiers, oscillators, and other circuits. Troubleshooting a dead system requires you to isolate the faulty circuit. Having done that, the defective component must be located.

There are several different ways to locate a faulty circuit. The best method is the one you like best, and the one that works best for you.

Cutting the Time Required for Signal Injection or Signal Tracing

System troubleshooting is sometimes done by starting at one end and working toward the other. At least, that is the procedure often described in books. Also, a system diagnostic (discussed in Chap. 6) usually starts at the input end of the system.

You can likely save time by starting in the *middle* of the system. An example will be given using the radio (see Fig. 2-24).

If you inject the signal at the center tap of the volume control and get an output signal at the speaker you have eliminated the possibility of trouble in all of the stages between the volume control and speaker. In other words, you have divided the receiver in half and reduced the time for troubleshooting.

For a quick injection method, tap the center tap of the volume control with a screwdriver and listen for a clicking noise in the output. Be sure the volume control has been set for high volume when you perform this test. Figure 5-1 shows this crude method of injecting a signal.

You cannot determine anything about the alignment by using a pulse or the screwdriver test. Some technicians use the blade of a screwdriver for signal injection when they are working away from their bench instruments. To demonstrate how wide a range of frequencies this can produce, try tapping the blade of a screwdriver onto the antenna terminals of a television receiver and note that flashes appear on the screen. You will also hear clicking and static noises in the speaker. Using a screwdriver blade takes a certain amount of skill and experience so that you know what to expect for the particular system you are servicing.

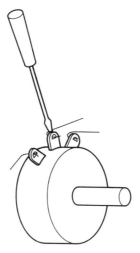

Figure 5-1 A quick injection method for troubleshooting.

This is not a discussion about how to use screwdrivers instead of test equipment. It is merely included as a time saver. Remember that the screwdriver injection method does not tell you anything about quality of the sound being passed through, or the amount of gain available from, each stage.

In both the signal injection and in the signal tracing method there should be a change in the amount of gain from the input of the amplifier to the output of the amplifier. Technicians train themselves to listen for this. If they see no change in the gain further investigation of the stage is recommended. However, keep in mind, though, that it *may* be normal that no stage gain is noted. An example would be in the case of an emitter follower stage.

Instead of using a screwdriver, you may prefer to use a signal injector. This simple device injects a pulse or square wave into the system being tested. Because of their broad harmonic content, signal injectors can be used in r-f, i-f, and audio stages.

Signal injectors can be purchased in kit form, or you can build one from scratch using the circuit described in Chap. 11.

After using the injection method to determine which half of the system is at fault, proceed with either the signal injection or signal tracing procedure.

Using a Radio as a Signal Injector

Not all of the circuits in a system are amplifiers. In a radio, for example, there is a local oscillator and a detector. In other systems there may be other nonamplifier circuits.

Since most of the techniques in this chapter are for amplifiers in systems, the methods of locating oscillator and detector faults will be treated separately here. Remember, these techniques are applicable to a variety of other systems.

Suppose the oscillator in the receiver of Fig. 5-2 is defective. One method to

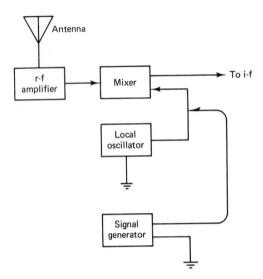

Figure 5-2 A block diagram of the receiver.

determine if it is working is to inject a signal generator signal at the proper frequency into the system.

In an AM receiver, like the one used as an example in this book, the oscillator frequency is always 455 kilohertz above the frequency being tuned. So, when they are subtracted in the converter (or mixer), the difference is the 455 kHz i-f frequency.

Example

A dead AM receiver with a defective oscillator circuit is tuned to a station at 1000 kHz. What oscillator frequency should be injected to make the receiver play?

Solution

$$\begin{aligned} \text{Station frequency} &= 1000 \text{ kHz} \\ + 455 \text{ kHz} \\ \hline \text{Oscillator frequency} &= 1455 \text{ kHz} \end{aligned}$$

This local oscillator signal should be injected by the signal generator into the converter or mixer stage.

The generator signal is injected at the same point as the output of the oscillator circuit. A capacitor is advised to isolate the generator from the receiver DC circuits.

NOTE

Most signal generators have a built-in isolation capacitor. Check to see if it is present in your signal generator. If so, do not add another isolation capacitor.

Figure 5-3 shows the point of oscillator injection for the receiver used as an example in this book.

A quick way to check the oscillator is to position another receiver back-to-back with receiver A, which is being serviced. Receiver B is the good receiver in the simplified test setup shown in Fig. 5-4. The volume control of receiver A is turned up, and it is to a strong local station. The volume control of receiver B is turned down all the way, and it is set to the same station. Under this condition, the local oscillator of receiver B will inject a radiated signal into receiver A. In most cases, receiver A will begin to play. You may have to move the tuning adjustment of the receiver (B) back and forth to use this trick. It is a simple, quick way to determine that the oscillator is working in the receiver being serviced.

Troubleshooting by Signal Injection

Signal injection is used to locate a defective amplifier in a dead system. The idea behind this method of testing, shown in Fig. 5-5, is to inject a signal into each amplifier—one at a time—and observe the output of the system.

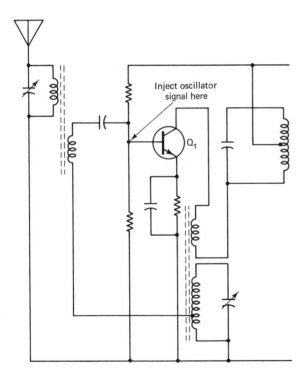

Inject oscillator signal here

Q_1

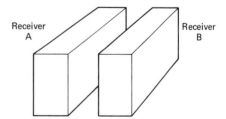

Receiver A   Receiver B

Figure 5-3   Injection of the oscillator signal directly into the circuit.

Figure 5-4   One method of checking the oscillator of a receiver is by using the oscillator in a receiver that is known to be working.

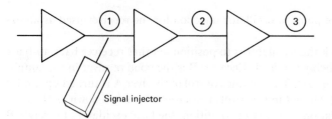

Signal injector

Figure 5-5   Troubleshooting by signal injection.

If you start by injecting the signal at 1, then 2, then 3, and so forth, you will not get an output signal until you have gone past the trouble. So, if you inject the signal at 2 and do not get an output, but inject it at 3 and do get an output, the trouble must be between 2 and 3.

There is an advantage of injecting the signal in the sequence just described. Each time you move the signal source the output signal will get weaker unless you increase the strength of the signal from the signal generator. This is better than starting at the output and moving back. Moving the other way requires a relatively strong signal to be injected at the output. When you move back to the previous amplifier, you can overdrive it enough to cause damage.

What if the amplifiers in the illustration are high-frequency tuned types? In that case, use a signal generator that is adjusted for the frequency that is supposed to pass through the amplifier stages. This will eliminate the possibility, although rare, that the trouble is in improper alignment of the tuned circuit between the amplifiers.

Signal Tracing

Instead of injecting the signal at each stage and monitoring the output, another technique is to inject a signal at the input and trace the path of the signal through the system. A good oscilloscope can be used to look at the signal at various points along the signal path. This is shown in Fig. 5-6.

Instead of an oscilloscope, a speaker or headset can be used to trace the signal through an audio system.

Special signal tracing equipment has been manufactured. This will be described in terms of a radio system, but the idea works with any kind of system. As shown in Fig. 5-7, the signal tracer actually consists of all of the stages that are available in a radio. So you probe various parts of the receiver with a switch delivering the signal to the corresponding part of the receiver that is in the signal tracer.

For example, suppose you are looking at the output of the detector. That goes to an audio voltage amplifier. So your signal tracer would be switched to deliver the signal from the output of the detector to the voltage amplifier of the signal tracer. If you get a signal there you know that everything from the antenna to the detector is working properly in the receiver being serviced.

In some systems this type of signal tracing equipment becomes very sophisticated and sometimes expensive. This is true for industrial electronic test stations

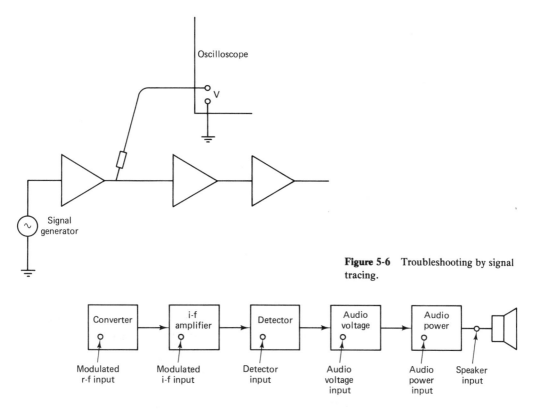

Figure 5-6 Troubleshooting by signal tracing.

Figure 5-7 The signal tracer is really a special form of radio.

where the equipment is specially designed and put in rack mounts. In some applications this technique is preferred over signal injection.

Signal Injection with the Amplifier Signal

Return again to Fig. 5-5. If you have reason to suspect one of the amplifiers, you can use a quick check to determine if that amplifier is defective.

For the purpose of this discussion we will assume that it was the amplifier between 2 and 3. A quick check to see if the amplifier is inoperative is to use a capacitor to connect the signal out of the first amplifier to the input of the third amplifier. This is shown in Fig. 5-8.

Of course, there will be some loss in the gain of the system because you don't have the advantage of the second amplifier. But it is quite often an easy matter to determine if an amplifier is not working by using this procedure. Naturally, you can't use a capacitor that won't pass a signal. For example, if this is an audio amplifier, you should use a 0.1-microfarad capacitor that has a relatively low reactance to the amplified audio signal. Otherwise, the capacitor will not be able to inject the signal.

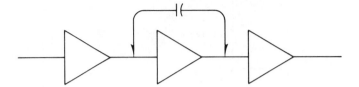

Figure 5-8   A quick check to see if an amplifier is defective.

Using the Proper Probe

Remember that if you are following an r-f signal, or looking for an injected signal in the r-f range, the oscilloscope may not be able to display it. Likewise, if you are looking for transient voltages, such as glitches, the scope may not be able to see them.

In the case of tests for analog amplifiers—where you are using a signal generator with a frequency too high for your scope—you can amplitude modulate the signal. This is done with the internal modulation feature on the signal generator. (If it is a function generator, you can add the audio modulation signal from an external point.) Then use a detector probe on the oscilloscope to demodulate the test signal. The oscilloscope will display the modulation signal, indicating that the r-f signal is passing through the amplifier. This works even though you can't see the r-f portion of the signal on the scope.

Using the Frequency Domain Display

You can get a good idea of the frequency response of an amplifier using the test setup shown in Fig. 5-9. There is an important restriction on this test. The output of the signal generator must be a constant value over the range of frequencies being tested.

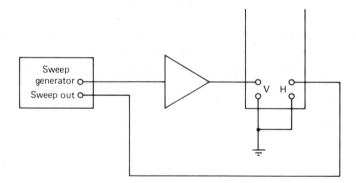

Figure 5-9   The variable DC voltage changes the frequency of the generator. At the same time it changes the position of the trace on the oscilloscope.

It may be necessary to monitor continuously the input amplitude to the amplifier. That will ensure that any changes in the output are due only to the amplifier and *not* to the changes in the signal generator output.

The test procedure is quite simple. The output of the amplifier is delivered to the vertical input to the scope. Then the signal generator is tuned through the range of amplifier frequencies. The amplifier output is monitored.

For this test it is a good idea to turn off the horizontal sweep of the scope because you are only looking at the amplitude here.

As you adjust the generator through the range of frequencies, you are making sure there is no radical change in the output amplitude. Such a change would indicate an uneven frequency response.

Remember that the bandwidth of the amplifier is the range of frequencies between seven-tenths of the maximum voltage amplitude.

It is absolutely necessary that the amplifier does not clip the waveform at any part of the test. This would be indicated by a bright spot on the end vertical line. If that happens, you have to reduce the input and rework the test. Remember, you are looking for undistorted output in the frequency range.

Using the Sweep Technique

Instead of using a DC voltage for getting the frequency domain display on the oscilloscope, as in the test just described, you can use a sweep generator or function generator with a VCO input. This test setup is shown in Fig. 5-10. The sine wave voltage

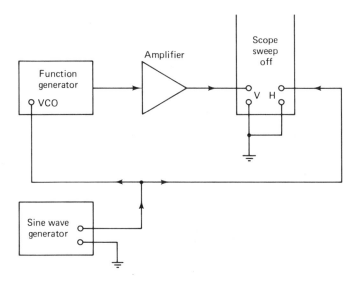

Figure 5-10 Frequency response of an amplifier can be determined with a sweep generator. The oscilloscope must be in the frequency domain mode.

causes the function generator (or sweep generator) to sweep back and forth throughout the range of frequencies that you are interested in.

The sine wave generator moves the oscilloscope sweep back and forth at the same rate that it moves the frequency of the generator up and down. So, when the sine wave is at its lowest amplitude, the generator is delivering its lowest frequency. Conversely, at the highest amplitude of the sine wave the generator is delivering its highest frequency.

One of the problems with the test setup so far is that you don't know just exactly what frequencies you are looking at. You can mark the gradicule division for corresponding frequencies throughout the range by starting with the test setup shown in the previous section. Using a DC voltage and a frequency counter, you can mark important points on the graticule corresponding to significant values of frequency.

As an example, suppose you are checking an audio amplifier to determine that it has the ability to pass all frequencies from zero to 100,000 hertz. Using the DC power supply you can determine what voltage is necessary to produce 100,000-H output. That would be the maximum positive *peak* voltage of the sine wave injection signal. Likewise, the lowest value would be the one corresponding to the function generator's lowest frequency (usually very near zero hertz).

Having marked this range of frequencies, you can also mark the center frequency. Figure 5-11 shows a marked graticule.

Now a sine wave is injected with the peak-to-peak signal corresponding to the range of DC voltages that you have just determined. The signal generator will sweep all of the frequencies when that sine wave voltage is applied to the VCO input.

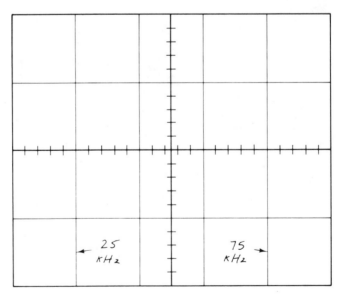

Figure 5-11 You can mark the graticule for specific frequencies on the frequency domain display.

Remember, a sine wave is not linear. So the frequencies at the high and low ends may tend to be crowded together. You can eliminate this crowding by using a sawtooth voltage for sweeping the frequency, but it isn't usually necessary to get the type of accuracy required for this setup.

When it is necessary, as when aligning an i-f stage, a sweep generator may have provisions for marker inputs. These markers are simply fixed frequencies that produce birdies on the trace. See Fig. 5-12. These birdies tell you exactly where the fre-

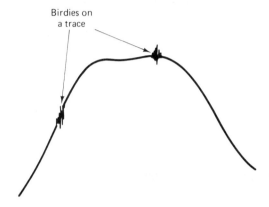

Birdies on
a trace

Figure 5-12   Marker birdies on the
frequency domain display.

quencies of interest are located. The manufacturer will tell you where to set these frequencies so that you can get the desired frequency response.

It requires a high-voltage injected signal, but the marker signals can be introduced into the Z axis of the scope. The Z axis is usually located on the back of the scope. It is also called intensity modulation. An input signal to the Z axis will cause a bright spot on the trace, so it is possible to use this terminal for introducing a marker.

In many cases the Z axis requires a relatively high voltage, as much as 35V. However, this method of marking has the advantage that it does not distort the frequency domain trace.

The Role of AFC and AGC

If you are going to perform *any* kind of sweep measurement in a system that has an automatic frequency control (AFC) or an automatic gain control (AGC/AVC), it is a very good idea to deactivate these circuits before you start measurements. The reason is that the automatic gain control tends to smooth out any changes in amplitude that would occur as a result of poor amplifier frequency response. Likewise, the automatic frequency control will cause the system under test to track the sweeping frequency. That, in turn, will make it impossible to tell where you are on the frequency domain display.

Deactivating these circuits is usually accomplished easily by the use of a DC power supply. A DC voltage is used to replace the DC output in the closed loop of either circuit. You may be able simply to clamp the AGC voltage by connecting a DC

source across it. In other cases it will be necessary to disconnect the AGC/AVC output. In any event it is worth the time that it takes to deactivate these circuits so that
you get a better presentation of the amplifier's response.

Noise Generators

You will remember that a sawtooth and a square wave contain a wide range of harmonic frequencies. That accounts for their usefulness in testing certain types of
circuits.

Another signal that has a very wide range of frequencies is *white noise.* White
noise gets its name because it supposedly contains all audio and r-f frequencies, as in
the case of white light that has all visible frequencies. In reality, this is not quite true,
but the tradition persists. Pink noise is a noise that has more lower frequencies than
higher frequencies.

One of the best sources of white noise is a semiconductor diode. Resistors also
produce a considerable amount of white noise in a circuit.

In the formal method of aligning a discriminator or ratio detector (two kinds of
detector circuits for FM signals), a sweep generator is used to sweep the range of frequencies and the output is displayed on an oscilloscope.

This procedure requires the same test setup as used for viewing the bandwidth of
an amplifier on a scope. A quick way to check the amplifier and detector circuits

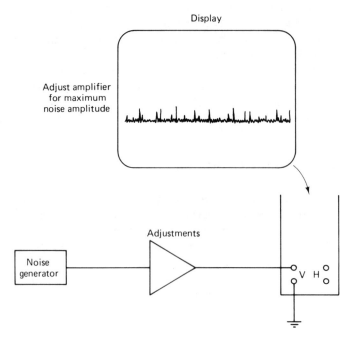

Figure 5-13 A noise generator can be used for quick alignment of bandpass amplifiers and detectors.

without resorting to the sweep method is to inject a white noise from a white noise generator into the circuit. See Fig. 5-13. With the noise injected, the circuit is adjusted for maximum output noise. When the noise is maximum the range of frequencies being passed is also maximum.

If you are going to develop this technique it is a good idea to compare the white noise input with the formal sweep method because it is a qualitative test and requires some experience in its use.

SUMMARY

Two very valuable methods of troubleshooting any electronic system are signal tracing and signal injection. The methods are similar. You are looking for a section in the system that will not pass the signal.

As a rule, the equipment needed for signal tracing is more complicated (and more expensive) than that for signal injection. You should be able to use either method in troubleshooting.

In this chapter, and elsewhere in the book, there have been discussions on how to make certain tests without using formal test equipment. *Do not be misled into thinking that test equipment is unnecessary!* While the quick tests are useful, the amount of information you get is very limited.

Using the oscilloscope in the frequency domain rather than the time domain is a very good way to check the frequency response of an amplifier. The manufacturer's literature will give specific details on how and where to inject signals. The slightly greater amount of time needed to set up a frequency domain test is well worth the effort.

Remember these precautions for frequency domain tests:

- Be sure to deactivate closed-loop circuits (such as AFC) before attempting a frequency domain test.
- Make sure the amplitude of the test signal is constant throughout the range of sweep frequencies.

PROGRAMMED SECTION

The instructions for this programmed section are given in Chap. 1.

---

1. Troubleshooting a system means to
   A. isolate the defective circuit. Go to block 7.
   B. find the faulty transistor. Go to block 14.

---

2. You have selected the wrong answer for the question in block 17. Read the question again. Then go to the block with the correct answer.

---

3. You have selected the wrong answer for the question in block 18. Read the question again. Then go to the block with the correct answer.

---

4. The correct answer to the question in block 12 is A. Despite the advantage, marking with the Z axis is not popular with technicians—probably because of the high-amplitude marker signal needed.

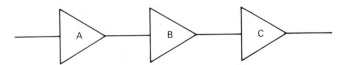

   Here is your next question:
   Amplifier B in the system shown in this block is suspected of being defective. There is no output from amplifier C. What single component could be used to determine if the fault is amplifier B? Go to block 21.

---

5. The correct answer to the question in block 6 is C. A modulated signal from the signal generator is used, and a demodulator probe is used with the oscilloscope. The audio modulating signal is displayed on the oscilloscope screen.
   Here is your next question:
   Which of the following scope terminals is used for intensity modulation of the trace?
   A. Sync input. Go to block 8.
   B. Z axis. Go to block 12.

---

6. The correct answer to the question in block 11 is B. The oscillator frequency is 455 kilohertz above the tuned frequency.
   Here is your next question:
   Using an oscilloscope, you are tracing an r-f signal. The oscilloscope cannot reproduce the r-f signal. You can see the signal by
   A. modulating the signal generator r-f and using a low-capacity scope probe. Go to block 13.

B. demodulating the signal generator r-f and using a demodulator scope probe. Go to block 5.

C. Neither choice is correct. Go to block 16.

---

7. The correct answer to the question in block 1 is A. An amplifier is an example of a circuit system. It is made of components like resistors, capacitors, inductors, and transistors. The faulty component may not be a transistor. After a preliminary investigation, the first step is to measure the power supply voltage. Then isolate the faulty circuit.

   Here is your next question:

   Why is it usually a good idea to start troubleshooting a system by injecting a signal in the middle of the system?

   A. to save time. Go to block 15.

   B. because that is the place where the trouble is usually located. Go to block 9.

---

8. You have selected the wrong answer for the question in block 16. Read the question again. Then go to the block with the correct answer.

---

9. You have selected the wrong answer for the question in block 7. Read the question again. Then go to the block with the correct answer.

---

10. You have selected the wrong answer for the question in block 15. Read the question again. Then go to the block with the correct answer.

---

11. The correct answer to the question in block 18 is B. By starting at the input and working toward the output, there is less chance of overdriving an amplifier. You have to continually increase the signal strength to get an output, when you go from a point of no output to a point where an output occurs.

   Here is your next question:

   The oscillator frequency of an AM radio is measured with a frequency counter. It is 1095 kilohertz. The radio is tuned to

   A. 1095 kHz. Go to block 20.

   B. 640 kHz. Go to block 6.

---

12. The correct answer to the question in block 5 is B. Another use for the Z axis is for marking the scope time base. A short-duration pulse at the Z axis produces bright spots. This procedure is shown in this block.

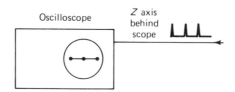

Here is your next question:

Which of the following is a reason for using the Z axis of the scope for a frequency marker?

A. It doesn't distort the pattern. Go to block 4.

B. It is more accurate. Go to block 19.

---

13. You have selected the wrong answer for the question in block 6. Read the question again. Then go to the block with the correct answer.

---

14. You have selected the wrong answer for the question in block 1. Read the question again. Then go to the block with the correct answer.

---

15. The correct answer to the question in block 7 is A. Dividing the system in half is usually a good idea for troubleshooting any system, not just for radios.

Here is your next question:

An advantage of using a signal injector over a screwdriver is that

A. it is a quicker method. Go to block 10.

B. the wide range of frequencies makes it useful for testing many types of analog systems. Go to block 17.

---

16. You have selected the wrong answer for the question in block 6. Read the question again. Then go to the block with the correct answer.

---

17. The correct answer to the question in block 15 is B. Keep in mind the fact that the systems in this chapter are assumed to be analog.

Here is your next question:

An AM radio is suspected of having a dead oscillator. It is tuned to a strong local station at a frequency of 1240 kHz. What signal generator frequency should be used to substitute for the oscillator output?

A. 1240 kHz. Go to block 2.

B. 1695 kHz. Go to block 18.

---

18. The correct answer to the question in block 17 is B.

$$(1240 + 455) \text{ kHz} = 1695 \text{ kHz}$$

Here is your next question:

There are two possible ways to use signal injection when troubleshooting a system. Which is usually the better way?

A. Start at the output with a strong signal and work toward the input. Go to block 3.

B. Start at the input with a weak input signal and work toward the output. Go to block 11.

---

19. You have selected the wrong answer for the question in block 12. Read the question again. Then go to the block with the correct answer.

20. You have selected the wrong answer for the question in block 11. Read the question again. Then go to the block with the correct answer.

21. The correct answer to the question in block 4 is capacitor. It is used to bridge the signal from the output of amplifier A to the input of amplifier C. There will be a loss of performance, but the system will work on a limited basis if the trouble is in amplifier B.

You have now completed the programmed section.

TEST YOUR KNOWLEDGE

1. Noise signals that cover the audio and r-f frequency ranges are called ____ noise.

2. Is the following statement true? When checking the frequency response of an amplifier, it is important to maintain the output of the amplifier constant throughout the test. ____

3. Name two types of signal generators that can be used for displaying the frequency response of an amplifier on a scope. ____, ____

4. What is a disadvantage of using a sine wave voltage for sweeping the oscilloscope trace during a check for frequency response? ____

5. When displaying the frequency response of an amplifier, the oscilloscope is set up for a ____ domain display.

6. What simple tool can be used for signal injection? ____

7. The intermediate frequency for FM broadcast radios is 10.7 megahertz. If a receiver is tuned to a station at 101.4 megahertz, the local oscillator frequency should be ____ megahertz.

8. Which of the following waveforms would be better for signal injection? ____
   A. sine wave
   B. pulse

9. After preliminary checks, for quick troubleshooting start ____.
   A. at the signal input end of the system.
   B. at the signal output end of the system.
   C. in the middle of the system.

10. Incorrect alignment is a ____.
    A. frequent cause of a system that doesn't work.
    B. rare cause of a system that doesn't work.

ANSWERS
TO TEST YOUR KNOWLEDGE

1. white

2. not true. It is necessary to keep the input to the amplifier at a constant amplitude. Some signal generators are designed to maintain a constant output amplitude. This is a very useful feature.

3. sweep generator, function generator with VCO input

4. crowding of the frequency at the upper and lower ends of the frequency range

5. frequency

6. screwdriver

7. 112.1

8. B. The high harmonic content of a pulse makes it useful over a wide range of frequencies.

9. C

10. B. Do not be in a hurry to align tuned circuits in a system. Only in rare cases would alignment be the cause of a system failure. (An exception is the tuned circuits in kit radios.) Think of alignment as a last resort in troubleshooting.

6

Symptom Analysis, Diagnostics, and Statistical Methods

CHAPTER OVERVIEW

The point has been made that it isn't likely that a single approach to troubleshooting and servicing equipment can be constructed. Technicians develop their own favorite techniques and, even though they may seem cumbersome to another technician, they probably work very well.

Regardless of how well your favorite method of troubleshooting works for you, sooner or later you hit that brick wall. It is the tough dog problem that seems to defy all reason.

When that happens, you need to convert to an alternate method. Here are some of the possibilities:

- Take another look at the symptoms. Is it possible that some other section could cause that symptom?
- Is it time to consider the shotgun method? (Replace every part—one at a time—in the circuit that you know is defective? There are cases where manufacturers actually recommend the shotgun approach.)
- What do the records say? Your own records may be helpful. Also, there are computer-based records maintained by manufacturers and private companies. So go for help.
- Is there a flowchart or diagnostic available for the system you are working with? Consult the literature to see what is available.

Sometimes it helps to set the big problem aside and come back to it a little later. In this chapter there is a brief look at some alternatives.

Objectives

Some of the topics discussed in this chapter are listed here:

- What happens if the symptoms don't help?
- What is a diagnostic?
- What are the meanings of the shapes in a flowchart?
- What is the order of statistical analysis?
- Can a new battery be at fault?

SYMPTOMS

If there is any such thing as a common denominator in troubleshooting it would be to start by observing the symptoms. Even that can be disputed. A shop manager may say "start by doing the paperwork." But, for the purpose of this discussion, we will presume that the paperwork has been done.

You cannot rely on a description of symptoms from the customer, or from a shop foreman, or from any source other than your own observation. After all, as a technician, you are a trained observer and you can recognize types of symptoms better than the untrained.

At one time the author of this book was a television technician in the field. A customer called to complain that he had been watching a football game and after the game he was not able to get the yard markers off the television screen. It turned out that the so-called yard markers were simply retrace lines.

This example may seem silly, but it tells you what can happen when customers (or any untrained personnel) describe symptoms for technical equipment they are not trained to troubleshoot. There have been attempts to set up complete troubleshooting procedures based on symptoms only. In some cases this has been successful. For example, in a radio broadcast transmitter, the symptoms of failure are often indicated by taking readings from the various meters.

Symptom analysis may not work in cases where there is a large turnover of designs, such as television receivers. For example, technicians are trained to observe that no sound and no picture usually mean trouble in the low-voltage power supply. That may seem like an obvious conclusion. After all, the sound section and the video section are both operated from the low-voltage supply.

However, if the low-voltage power supply in a receiver gets its energy from the flyback transformer, the trouble could be in the horizontal oscillator stage. In that case the memorized symptom would not be very useful.

As a matter of fact, for every new design there are some possible new symptoms that would need to be learned for effective troubleshooting. So symptom analysis is very useful, but not as the only method of troubleshooting electronic systems.

As demonstrated by the example of the low-voltage power supply, it should be remembered that your knowledge of the system is very useful. Where there is no sound and no picture and no brightness, the question is: What part of the system supplies both the sound and video? In that case, knowledge of the system leads to the most obvious conclusion that it is a low-voltage supply or at least a point where DC voltages apply to both circuits.

Open the Cabinet

So far, none of the techniques has required the technician to touch the equipment. It has all been a mental exercise. Sooner or later it is necessary to open the cabinet and make some measurements and decisions based on those measurements. If you are talking about a first step in a hands-on troubleshooting procedure that first step should always be: *Measure the power supply voltage.* This has been repeated over and over in this book because it is very important. Nothing happens in electronic systems unless the power supply voltage is present.

If the voltage *is* present, but it is not the correct value, it can produce apparent troubles in any section of the system. As an example, suppose the audio section of a radio has a transistor-sensitive amplifier—that is, an amplifier that produces a great amount of distortion if the supply voltage is not correct. The symptom would be *audio distortion.* The real problem would be in the power supply.

After measuring the power supply voltage and determining that it is correct, the next step depends greatly on the type of system involved. For example, in repairing a radio, it is necessary to locate the troubled section. Isolate the defective component and replace it within a few minutes. Otherwise, it would be cheaper to buy the customer a new radio than to suffer the financial loss that would be incurred by a technician spending too much time on a system that cost less than $100.

In a computer the next step would be to run a diagnostic that is on a floppy disk. In a tape recorder you can use visual observations to see if the tape is moving through the system and if it appears to be moving at the right speed. These are valuable as a next step.

It is a good idea to look for hot spots, burn marks, and places where a voltage arc over may have occurred. All of these things are a part of your general routine when you start a troubleshooting procedure.

THE DIAGNOSTIC

A diagnostic is a step-by-step procedure for troubleshooting an electronics system. Diagnostics take on several forms. One is the simple *flowchart.* Symbols that are often used on flowcharts are shown in Fig. 6-1. A typical flowchart is shown in Fig. 6-2.

The manufacturer leads the technician through a step-by-step procedure for iso-

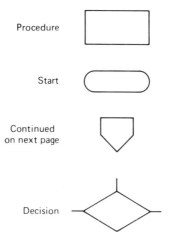

Procedure

Start

Continued
on next page

Decision

Figure 6-1 Symbols used in flowcharts.

lating troubles in a television set. This diagnostic is typical. Note that the diamond-shaped boxes in the diagnostic represent a decision to be made, and the square blocks represent individual steps in the troubleshooting procedure.

This type of troubleshooting is very effective, but it has one drawback. If you have to start at the beginning every time you troubleshoot a piece of equipment, you are going to lose some time on problems that don't occur until late in the troubleshooting diagnostic procedure.

In other words, you can often save time in some procedures by jumping ahead *if the symptoms indicate trouble in a specific section.*

Many technicians have reported that they run through the complete diagnostic procedure only once. From that time on they are sufficiently familiar with the equipment to troubleshoot it without resort to flowcharts. However, if you are trying to fix one of those tough dogs, it would be a good idea to start at the beginning. If you are not sure where the trouble *is,* you can't be sure where it *isn't.*

Experienced technicians claim that one of the most useful tools they have is the notebook they carry. They make notes on troubles that they have found in specific kinds of equipment. Their experience has been that troubles repeat. If you haven't seen a specific VCR for six months, it is easy to forget the cause of a particular symptom. So, after they locate it they jot it down in a notebook. Technicians claim that their notebooks are one of their most useful tools. This is always a good idea, but, it requires a certain amount of self-discipline.

There is another good reason for keeping a personal notebook. Manufacturers may deny this, but there are weaknesses in many models of electronic systems. For example, an SCR may burn out frequently in one design, a power transistor may burn out often in another design, and so on.

When several technicians are working for the same company and on the same equipment, it is a good idea to compare notes. Also, keep an up-to-date file on company change notices.

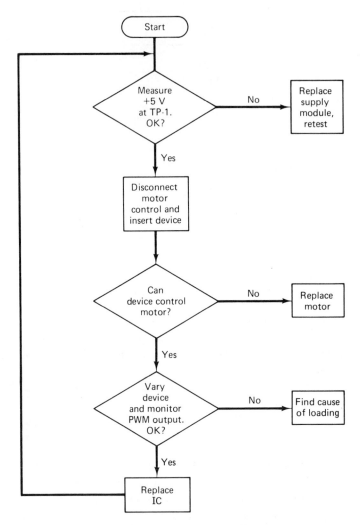

Figure 6-2   A typical flowchart.

Some companies provide a computer analysis for some types of equipment. For example, if you are troubleshooting television receivers you can subscribe to a computer system that tells you the most likely components for a specific type of failure. Subscription to this type of system can save many hours of frustration and will certainly repay in terms of efficiency.

If you repair computers you are certainly familiar with diagnostic programs that run the computer through its paces. These programs will tell you what specific section has failed in the diagnostic procedure. Diagnostics are well worth their cost and they are readily available.

Statistical Analysis

If you get to the point where you feel it is necessary to shotgun a circuit, it's smart to consider statistical analysis. With this approach, you replace components in the order that they are most likely to fail.

Statistical analysis is not just for shotgunning a circuit. Many times the trouble will be traced to a specific circuit, and there are two or more components that can produce the symptom.

Consider the power amplifier circuit in Fig. 6-3. It is conducting at near saturation. Here are some possibilities:

- There are shorted turns in the inductive output.
- Resistor R_1 is greatly reduced in value.
- Resistor R_2 is greatly increased in value.
- The transistor has an emitter-collector short circuit.
- Capacitor C_1 is shorted or very leaky.

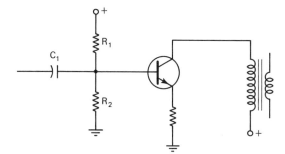

Figure 6-3 A power amplifier used in the discussion of troubleshooting.

If all of these components are soldered to the printed circuit board, you have to make a decision which to change first. This assumes that you don't have an in situ tester like the one described in Chap. 11. Even if you have an in situ tester, you have to make a decision about which remaining component is most likely to be the cause of the trouble.

Table 6-1 lists the components in the order of their most likely failure.

For the example of the power amplifier, the most likely component to fail is the power transistor. A common fault with bipolar power transistors is an emitter-collector short circuit.

Unnecessary soldering and unsoldering components on a printed circuit board is a good way to introduce new troubles. Certainly, if the capacitor is defective, the voltage at the base of the transistor will not be the correct value. So, before unsoldering any component in the circuit of Fig. 6-1, measure the emitter-to-base voltage and also the base voltage of the transistor.

If the manufacturer's schematic doesn't tell you what that voltage should be, you can make a quick estimate using the proportional method for finding voltages. This is

TABLE 6-1 STATISTICAL PROBABILITY OF FAILURE

Batteries

High-power devices
Does it get hot in normal use?
    Diodes, transistors, ICs, tubes
    Ultrasmall electrolytics

Other transistors, ICs, diodes

High-power passive devices

Low-power passive devices
    Capacitors
    Resistors
    Inductors

Based on information supplied by the Electronic Industries
Association.

an approximation, but it should be sufficient to tell you if a leaky capacitor is a likelihood.

The emitter-to-base short test for the transistor will tell you if the transistor is shorted internally.

To summarize, it is important to use measurements to determine the most likely component to fail. Having done that, and assuming the measurements have failed to identify the specific trouble, you can use Table 6-1 to set the order of part replacement. You should modify that table so that it corresponds with your experience from working on specific systems.

A very valuable troubleshooting tool is the in situ tester that makes it possible to check the electrolytic capacitor and/or the transistor while it is still in the circuit. These in situ testers are readily available.

As pointed out before, unsoldering components on a printed circuit board is not a good idea if you want to avoid working on the same system again in the very near future. In the system of Fig. 1-10, the most likely problem would be the battery that supplies the voltage to the system. Many times the person squawking the equipment will say, "It can't be the battery, I just put in a new one."

Unfortunately, new batteries can be defective.

Another problem you have to watch for is a customer or untrained personnel putting the battery in backwards. That works OK for flashlights (sometimes), but it can be highly destructive to transistor equipment.

SUMMARY

If you are troubleshooting familiar equipment, you will seldom have to refer to a flowchart or other kind of diagnostic. Even the tough dogs can be analyzed using standard troubleshooting procedures.

When everything else fails to get to the source of the trouble, it is best to turn to the manufacturer's flowcharts and other aids (such as change notices and the shotgun method).

As a rule, the shotgun method is the absolute last resort. For large complicated systems, such as a personal computer, the diagnostic on a disk is the place to start. These disks run the computer through a series of programs designed to check every section.

Technicians claim that one of their best aids is their own personal notebook. They use the notebook to write down peculiarities of different models of equipment they have serviced.

Special helpful hints can also be recorded in a personal notebook. They are printed in company bulletins, technical magazines, and publications by technician organizations. Some companies provide computer access to data banks that store symptoms and cures for equipment problems.

A disadvantage of using a diagnostic or flowchart is that you have to start at the beginning. The problem may not be anywhere near the starting point.

If you have to make a decision between components that may be causing the problem, remember the statistical chart. Replace the component most likely to fail.

PROGRAMMED SECTION

Instructions for this programmed section are given in Chap. 1.

1. The schematic in this block shows a closed-loop analog voltage-regulated power supply. The series resistance sense network (R_1, R_2, and R_3) is connected across the output load resistance (R_L). The objective of the supply is to maintain the voltage across R_L at a constant value. Any change in that voltage will be sensed by the arm of variable resistor R_2 and delivered to the base of voltage amplifier Q_1. The emitter of Q_1 is held at a constant voltage by the zener diode.

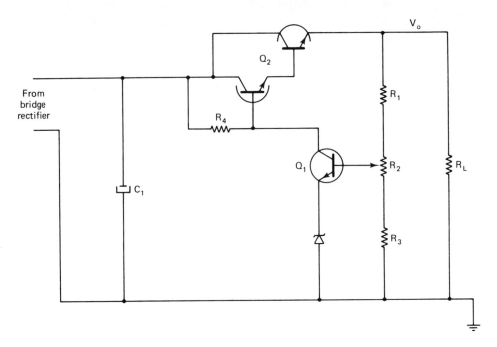

Transistor Q_1 conducts through R_4 and sets the bias for the Darlington power amplifier.

Suppose, as an example, the output voltage across R_L starts to rise. That causes the base voltage of Q_1 (picked off at R_2) to go in a positive direction. Transistor Q_1 conducts harder, and that produces a greater voltage drop across R_4. Increasing the drop across R_4 makes the base of the Darlington pair less positive, so the Darlington does not conduct as hard. Since the current through R_L (and the voltage divider) is now lower, the output voltage (V_o) will also be lower.

You are troubleshooting this power supply. The problem is that there is no change in output voltage when the variable resistor is adjusted. In normal operation R_2 is adjusted for the desired output voltage. The next step is to

    A. replace the Darlington pair because power amplifiers often go bad. Go to block 7.

    B. measure the collector voltage of Q_1 while adjusting the variable resistor. Go to block 14.

2. The correct answer to the question in block 5 is B. If the radio uses bipolar transistors the gain can be controlled by making the base either more positive or more negative.

    Sometimes the AGC voltage is delivered to the emitter (or source, or cathode), so you must analyze the circuit to determine if the AGC voltage should go more positive or more negative.

    Here is your next question:

A certain radio makes a loud hissing noise and the sound is weak. Look for noise problems at

    A. the input end of the system. Go to block 16.

    B. the output end of the system. Go to block 8.

3. The correct answer to the question in block 11 is B. A leaky capacitor causes current to flow through R_2 and drops the output voltage. Of the choices given, Table 6-1 shows that C_2 is the most likely cause of the trouble.

    Here is your next question:

What does a diamond shape mean in a flowchart? Go to block 28.

4. You have selected the wrong answer for the question in block 13. Read the question again. Then go to the block with the correct answer.

5. The correct answer to the question in block 12 is C. If R is a low value (below 150 ohms) the circuit is a decoupling filter. It is used to prevent variations in the positive line voltage from affecting the amplifier.

    If R is a high value (at least one-third the value of R_L) the circuit is for low-frequency compensation. At low frequencies the transistor sees $R + R_L$ as a load resistance, so its gain is high. At high frequencies C is a signal short circuit to the transitor and sees only R_L as a load resistance. That makes its gain lower.

    The higher gain at low frequencies is needed to overcome the higher loss in the coupling capacitor.

    Here is your next question:

A flowchart for repairing a certain radio asks: Does the AGC voltage become more positive when the radio is tuned from no station to a strong local station? Is this an error in the flowchart?

    A. Yes. The AGC voltage should become more negative. Go to block 21.

    B. No. There is no error. Go to block 2.

6. You have selected the wrong answer for the question in block 17. Read the question again. Then go to the block with the correct answer.

7. You have selected the wrong answer for the question in block 1. Read the question again. Then go to the block with the correct answer.

8. You have selected the wrong answer for the question in block 2. Read the question again. Then go to the block with the correct answer.

9. You have selected the wrong answer for the question in block 15. Read the question again. Then go to the block with the correct answer.

10. You have selected the wrong answer for the question in block 22. Read the question again. Then go to the block with the correct answer.

11. The correct answer to the question in block 16 is B. Do not confuse noise and distortion. With a poor antenna system the problem would be noise. It would make a hissing sound. Distortion means you can't hear the sounds clearly. It happens when the audio amplifier is not properly biased so that part of the signal is being clipped or cut off.

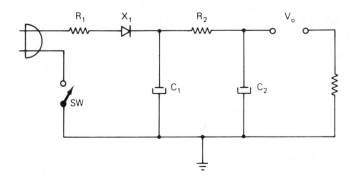

Here is your next question:
Refer to the half-wave rectifier in this block. The output voltage (Vo) is one-half the normal value without R_L connected. The diode is known to be good.

A. The most likely cause is R_1. Go to block 26.

B. The most likely cause is C_2. Go to block 3.

C. The most likely cause is R_2. Go to block 20.

12. The correct answer to the question in block 17 is A. If the capacitor is leaky, the voltage on the base of Q_2 will be too positive. The resulting increase in base current results in a collector current that is too high, so the voltage drop across the collector resistor will be too high. The collector voltage equals the power supply voltage minus the (too-high) voltage drop across the collector resistor and the collector voltage will be too low.

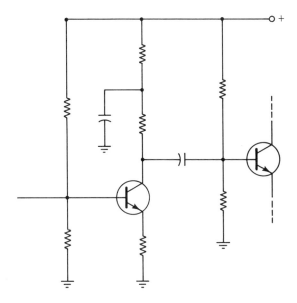

Here is your next question:

For the R-C-coupled amplifier in the circuit in this block

A. R and C form a low-frequency compensating network. Go to block 18.

B. R and C form a decoupling filter. Go to block 23.

C. It is not possible to choose the correct answer. Go to block 5.

---

13. The correct answer to the question in block 15 is B. (Refer to Table 6-1.) As a general rule, the hotter the device gets in its normal operation the more likely it is to become defective.

Here is your next question:

A program on a floppy disk that steps a computer through its operations is called

A. a flowchart. Go to block 4.

B. a diagnostic. Go to block 22.

---

14. The correct answer to the question in block 1 is B. Never start unsoldering and replacing components until you have made tests and measurements that can help you to zero in on the trouble. If the collector voltage of Q_1 does *not* change when you adjust R_2, then Q_1 is the likely problem.

Here is your next question:

Darlington amplifiers, like the one used for a power amplifier in the power supply of block 1, have a high power gain and a good voltage gain. The disadvantage of this circuit is that

A. it generates a lot of electrical noise. Go to block 24.

B. it generates internal heat. Go to block 15.

15. The correct answer to the question in block 14 is B. Generation of internal heat is one reason Darlington pairs are often used with heat sinks.

    Here is your next question:

Which of the following is more likely to be faulty in a circuit that is not operating properly?

A. a voltage amplifier. Go to block 9.

B. a power amplifier. Go to block 13.

---

16. The correct answer to the question in block 2 is A. In a system such as a radio almost all of the noise is generated in the first two sections.

    Here is your next question:

Which of the following can cause distorted audio in a radio?

A. poor antenna system. Go to block 27.

B. incorrect bias. Go to block 11.

---

17. The correct answer to the question in block 19 is B. There are test instruments that are especially designed for in situ tests.

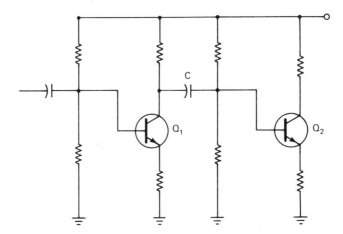

    Here is your next question:

Consider the R-C-coupled amplifiers shown in this block. If coupling capacitor C is leaky, you would expect the collector voltage of Q_2 to be

A. too low. Go to block 12.

B. too high. Go to block 6.

---

18. You have selected the wrong answer for the question in block 12. Read the question again. Then go to the block with the correct answer.

---

19. The correct answer to the question in block 22 is B. Sometimes you can start using a flowchart at an advanced point, but you have to be careful not to skip points that may indirectly lead to the trouble.

Here is your next question:

An in situ test is a test

A. that uses only a diagnostic approach. Go to block 25.

B. of a component while it is still in the circuit. Go to block 17.

---

20. You have selected the wrong answer for the question in block 11. Read the question again. Then go to the block with the correct answer.

---

21. You have selected the wrong answer for the question in block 5. Read the question again. Then go to the block with the correct answer.

---

22. The correct answer to the question in block 13 is B. Think of a diagnostic as being a step-by-step procedure for finding trouble in an electronic system. For computers this procedure is on a floppy disk. It is a program that steps the computer through a routine. A flowchart is another type of diagnostic approach. It is written rather than being on a floppy disk.

Here is your next question:

A disadvantage of using a diagnostic is that

A. it usually misses some very important things. Go to block 10.

B. it often starts you at the beginning of a test series, even though the symptom indicates that the fault is in a certain part of the system. Go to block 19.

---

23. You have selected the wrong answer for the question in block 12. Read the question again. Then go to the block with the correct answer.

---

24. You have selected the wrong answer for the question in block 14. Read the question again. Then go to the block with the correct answer.

---

25. You have selected the wrong answer for the question in block 19. Read the question again. Then go to the block with the correct answer.

---

26. You have selected the wrong answer for the question in block 11. Read the question again. Then go to the block with the correct answer.

---

27. You have selected the wrong answer for the question in block 16. Read the question again. Then go to the block with the correct answer.

---

28. The correct answer to the question in block 3 is that a diamond shape indicates a decision is to be made.

You have now completed the programmed section.

TEST YOUR KNOWLEDGE

1. Is the following statement true? The best information a technician can use for troubleshooting is an eye-witness account of what happened when the system failed. ____

2. Is the following statement correct? Symptom analysis will always work as a method of troubleshooting. ____

3. If there is no sound and no picture and no screen brightness for a defective television receiver, a likely place to start would be ____.
   A. the low-voltage power supply.
   B. the sync separator.

4. What is the first step in hands-on troubleshooting? ____

5. A certain radio has a power supply voltage that is too low. Can this cause a symptom of sound distortion? ____

6. What visual observations are useful when you first start troubleshooting a defective system? ____

7. Is the following statement correct? Sometimes there are weaknesses in the design of some equipment that cause repeated failures in certain models of equipment. ____

8. Which of the following components is more likely to fail in a system? ____
   A. resistor
   B. power transistor

9. Which of the following is more likely to fail? ____
   A. rectifier diode
   B. film-type capacitor

10. Is the following statement correct? The customer reports that new batteries have just been installed, so you should not waste time checking out the battery supply. ____

ANSWERS
TO TEST YOUR KNOWLEDGE

1. no, the statement is not true. Eye-witness accounts can be misleading and a waste of time. Use them as a *possible* aid in troubleshooting.

2. no, the statement is not correct. There are cases where there is trouble but no observable symptoms.

3. A

4. Measure the power supply voltage.

5. Yes. You should remember that a low power supply voltage may cause the weakest circuit in the system. In the case mentioned here that would be the audio amplifier.

6. Look for hot spots, burn marks, and places where a voltage arc over may have occurred.

7. yes, the statement is correct. These weaknesses are sometimes mentioned in service bulletins. You should put them in your notebook for future reference.

8. B

9. A. The rectifier diode normally gets warm (or hot) during its normal operation. Suspect hotter components before those that do not get hot.

10. no, the statement is not correct. It is best not to take the customer's word as final for any part of the troubleshooting procedure.

7

Servicing Closed-Loop Circuits

If the troubleshooting jobs are listed according to difficulty, then catastrophic failures would probably be the least difficult to service. Closed-loop circuits and intermittents are among the most difficult.

Closed-loop circuits can be put into two main categories: those that control voltage and those that control frequency (or phase).

Examples of *closed-loop circuits that control voltage* are AGC/AVC and regulated power supply. Examples of *closed-loop circuits that control frequency* are automatic frequency control and phase-locked loop (used as a frequency synthesizer).

Closed-loop circuits are difficult to service because their output depends on their input, and their input depends on their output. Any trouble anywhere in the circuit can change all of the voltages, currents, and frequencies in the circuit.

Fortunately, there is a standard procedure that works very well for both kinds of closed-loop systems.

Objectives

Some of the topics discussed in this chapter are listed here:

- What is the standard procedure for troubleshooting all feedback (closed-loop) circuits?
- How is frequency synthesis accomplished?

- What kind of voltage is usually used in both voltage and frequency closed-loop systems?
- How does a closed-loop circuit affect the alignment of a receiver system?
- Why isn't it correct to align a radio for maximum output?
- What is the disadvantage of an analog voltage regulator?

CLOSED-LOOP CIRCUITS FOR VOLTAGE CONTROL

The closed-loop circuits that control voltage will be taken up first. Figure 1-20 shows the block diagram of a simple radio. An important closed-loop circuit in this system is the *automatic volume control* (AVC). It samples the output signal voltage at the detector, filters it, and delivers a DC gain-control voltage to early stages such as the r-f or first i-f amplifiers. In other receiver systems, it is called automatic gain control (AGC). We will refer to it as the AVC/AGC circuit.

The purpose of this circuit is to keep the output sound signal—at the speaker—relatively constant, even though the strength of the incoming signal may vary. Such variations may be caused by changes in atmospheric conditions and changes in the height of the ion layer that affects long-distance reception.

The operation of the circuit is easy to understand. Suppose the input signal from the antenna increases in amplitude. That would normally cause an increase in sound volume from the speaker. However, part of the signal is rectified and filtered at the detector. The resulting DC voltage is fed back to the r-f and i-f circuits to reduce the gain of those stages. So the increase in sound volume at the speaker has been prevented.

Suppose, on the other hand, that the input signal from the antenna decreases. That would normally decrease the sound volume at the speaker. However, the AVC/AGC DC control voltage now increases the gain of the r-f and i-f amplifiers in order to keep the sound volume at the speaker constant.

So the AVC/AGC circuit produces a DC control voltage that holds the output sound volume constant by adjusting the gain of the r-f and i-f amplifiers.

As with all closed-loop systems, the difficulty in troubleshooting this one is the fact that the output (from the detector) depends on the amplification of the input signal, and the amount of a signal input amplification depends on the amount of detector output signal. If there is trouble anywhere in the loop it is difficult to find because everything depends on everything else within the loop.

There is only one way to troubleshoot this circuit effectively: *Open the loop*. This can be done several ways.

One way is to disconnect some part of the feedback loop and insert a DC power supply that substitutes for the AVC/AGC DC voltage. Another way is to clamp the DC control voltage by simply placing a DC supply across the automatic volume control circuit. Both methods are illustrated in Fig. 7-1.

In either case, the gain of the r-f and/or i-f amplifiers is controlled by the DC

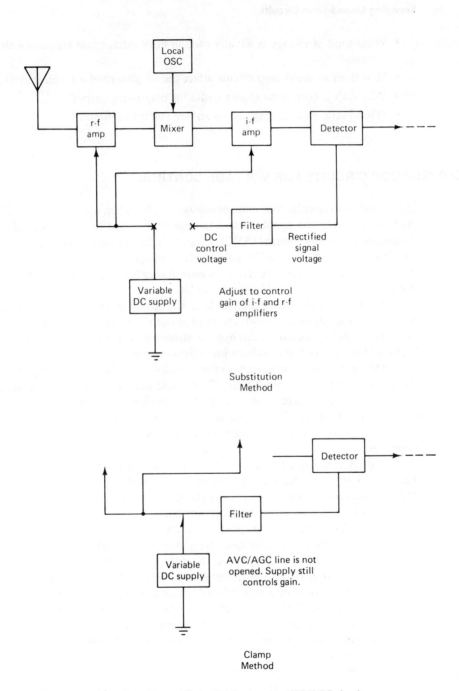

Figure 7-1   Two methods of defeating the AVC/AGC circuit.

power supply. Once this is done, the standard troubleshooting procedures can be used.

Clamping the AVC line (rather than opening the circuit and substituting a DC voltage) has the advantage of being quicker. It is not necessary to unsolder and re-solder, and that makes it a more reliable method.

Opening the loop has the advantage of making it possible to perform an important test of the circuit operation. By adjusting the power supply voltage in the opened-loop method, you are able to control the gain of the amplifiers. That, in turn, *should* change the amount of DC voltage available for feedback. If it doesn't, then the next step is to go to the output of the detector. Using an oscilloscope, look at the variation in detector input when you change the gain of the r-f or i-f amplifiers. If there is no change in the gain of the system when you vary the substitute DC voltage, then it follows that at least one of the amplifiers is defective.

The test setup in Fig. 7-1 is important for another reason. In some rare cases it is necessary to adjust or align the various r-f and i-f stages in a receiver. A common practice is to connect an AC voltmeter across the output in place of the speaker. See Fig. 7-2. Then the adjustments are made for maximum output voltage as indicated by

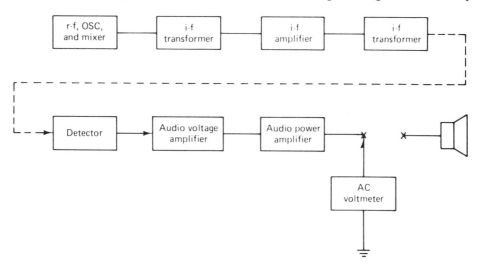

Figure 7-2 A measurement method sometimes used during alignment procedure.

the meter. This procedure is actually given with some radio kits to enable the kit builder to adjust the receiver for maximum number of stations.

There are several problems with this procedure. The automatic volume control (or automatic gain control) affects the output.

Suppose, for example, you are adjusting a transformer between the i-f stages. This adjustment, called *alignment*, is used to allow the receiver to pass only the i-f frequencies and reject all others. As the alignment of the first i-f transformer takes place, the amount of signal delivered to the i-f amplifier is increased or decreased.

Suppose the alignment procedure produces an increase. That also produces an increase in the AVC/AGC feedback loop *and reduces the gain of the receiver.* Hence, increasing the gain of the amplifier by the adjustment decreases the gain of the receiver by the AVC/AGC action.

The overall effect of the AVC/AGC, then, is to cause a very broad adjustment of the transformer and an improper indication at the output using the method in Fig. 7-2.

The second thing wrong with measuring the output audio during alignment as shown in Fig. 7-2 is that the *maximum gain usually does not correspond with the maximum fidelity.* To understand why, look at the illustration in Fig. 7-3. Note that

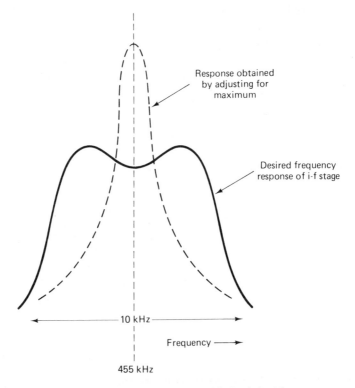

Figure 7-3 Comparison of maximum response with the desired frequency response of an i-f stage.

maximum signal amplitude passed through the transformer results in sharp skirt selectivity. So the range of frequencies, which in an AM radio should be about ±10 kHz, has been seriously decreased.

The effect, then, of increasing the output as measured on the meter is to decrease the fidelity of the receiver because *all of the useful information in an AM signal is in the sidebands.*

Keep in mind that these same facts are applicable to FM receivers, communications receivers, television receivers, and the like. Their message is clear: Defeat the AVC/AGC system before any alignment is attempted.

Frequency Domain

The most effective way to align the i-f stage or any other passband circuitry is to put the oscilloscope and signal generator into a frequency domain display and watch the passband while the adjustments are being made. This procedure, shown in Fig. 7-4, also works best with the AVC/AGC circuit defeated.

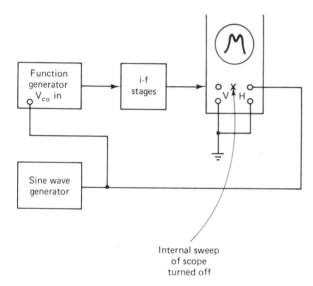

Internal sweep
of scope
turned off

Figure 7-4 Sweep method for looking at the i-f response curve on an oscilloscope screen.

The AVC/AGC Bias Voltage

Alignment can be performed while looking at the AVC voltage, but this has no greater advantage over looking at the audio output at the speaker.

The bias voltage used to replace the AVC/AGC should be a pure DC. You can get close to this by using a variable 5V supply, as shown in Fig. 7-5. The problem is that there may be some ripple getting through the supply. Also, any noise on the line can pass through this type of setup. The overall effect of ripple or noise is to upset the alignment procedure.

A better way is to use a battery pack, like the one shown in Fig. 7-6. At one time this had the disadvantage of being greatly dependent on the age of the batteries. So, over a long period of time, batteries had to be replaced. This was important because it could be a long time between uses. Today, the use of rechargeable Nicad batteries just about eliminates the problem of battery aging. Batteries can be periodically recharged as part of the calibration checkout period for the bench.

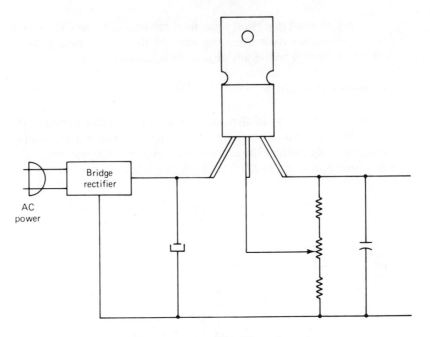

Figure 7-5   A variable 5V supply.

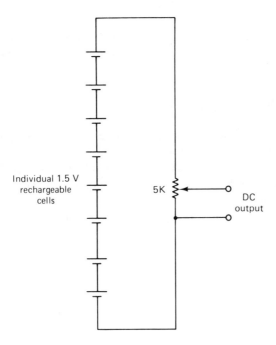

Individual 1.5 V
rechargeable
cells

5K

DC
output

Figure 7-6   Details of a battery
pack circuit.

A very important advantage of the battery pack used for bias substitution is that it has a floating ground. In other words, there is no fixed ground for the output voltage. That means that it can be connected across any of the circuits without fear of grounding the circuit through a common connection in the power supply.

The DC voltage used to substitute for the AVC/AGC voltage may need to be either positive, negative, or both positive and negative. Remember that the gain of a bipolar transistor amplifier can be controlled in two ways: forward AVC/AGC and reverse AVC/AGC.

In some complicated communications systems, the bias may be both forward (positive-going) and reverse (negative-going). The reason is that it increases the system stability. For example, if the r-f amplifier bias is negative-going and the first i-f amplifier bias AGC is positive-going, then the overall stability of the receiver is increased.

AVC/AGC systems are not only used in receivers. They are also used in hearing aids, public address systems, high fidelity audio systems, and industrial electronic systems. Regardless of where you encounter this type of gain control, the message is clear: You cannot do an effective job of troubleshooting or aligning them unless you defeat the closed-loop circuit.

Analog (Linear) Regulators

In Chap. 6 a simple closed-loop regulator was discussed (see block 1 of the Programmed Section). Figure 7-7 shows a block diagram of that system.

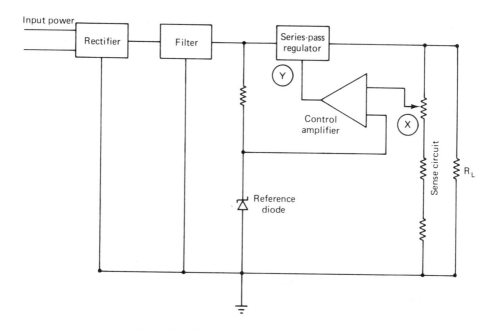

Figure 7-7   Block diagram of an analog regulator.

If you are going to troubleshoot this system, and the obvious preliminary methods don't work, open the loop and substitute a DC voltage at some point in the feedback loop.

The logical place to open the loop is at the lead to the adjustable arm in the variable resistor (the point is marked with an X in the block diagram). This is the center lead on the variable resistor. Unsoldering this lead *may* not involve unsoldering a connection to the printed circuit board.

If the variable resistor is mounted to and soldered to the printed circuit board, the next most logical place to open the loop is at the base of the power amplifier. That point (marked with a Y) is easy to reach and is not usually connected directly to the board.

So a new dimension is added to troubleshooting. In addition to the statistical method of choosing a possible faulty component, you must pick components that can be investigated with a minimum amount of physical damage to the circuit.

Figure 7-8 shows a circuit using a three-legged integrated circuit (IC) voltage-

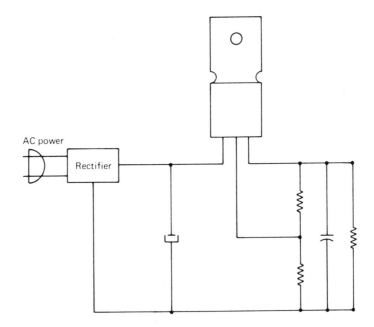

Figure 7-8   A fixed-voltage integrated circuit regulated supply.

regulated supply. These ICs are very popular. The closed-loop circuitry is inside the package. So, if your troubleshooting procedure points to this device, it is a replacement job.

The indications of trouble in the three-legged IC are overheating, input and output voltages equal, zero output voltage, and no supply regulation. Ripple in the output can be an indication of a faulty filter capacitor or a bad regulator.

Switching Regulators

A disadvantage of a linear closed-loop regulator is that it is not efficient. A better way is to use a switching regulator—a better way, that is, if the most important consideration is efficiency.

Figure 7-9 shows the block diagram for a switching regulator. Here you see that

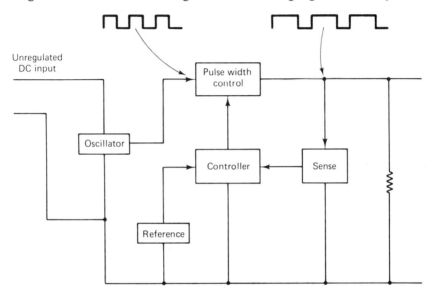

Figure 7-9   Block diagram of a switching regulator.

a relaxation oscillator supplies a pulse signal through the power control amplifier. Regulation of the output is obtained by feeding back part of the output voltage and using it to control the width of the pulses.

Wider pulses represent a high output power and corresponding higher output voltage.

Because the frequency is high, the pulses are very easy to filter. So less expensive and more effective filtering can be used in this closed-loop regulator.

To troubleshoot, start by opening the loop and substituting a DC control voltage. An oscilloscope is useful for troubleshooting this type of circuit. You can use it to look at the output waveform of the oscillator to be sure it is there, and that there is a change in pulse width corresponding to change in the DC substitution voltage.

If you can't get the variation in pulse width, there is something wrong with the switching part of the circuit. If you can't get any output from the supply, you have probably already discovered that the oscillator isn't working.

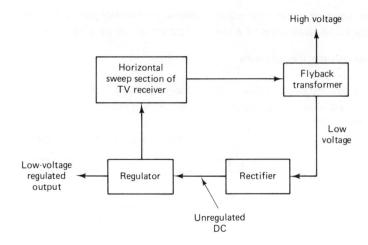

Figure 7-10   Block diagram of a scan-derived power supply.

A specialized regulated power sometimes used in television receivers is illustrated in Fig. 7-10. It is called a scan-derived supply. It gets its input power from a winding on the high-voltage (flyback) transformer. The flyback transformer is energized by pulses from the horizontal oscillator, and the horizontal oscillator gets its energy from the scan-derived supply.

When this system is first energized, nothing can happen. The horizontal oscillator cannot oscillate because it is not receiving any power. So there is no AC power delivered to the transformer and the regulator does not receive any power.

This deadlock is broken by a *start-up circuit*. It is simply an oscillator that starts the circuit operation. After the circuit is started into operation, the start-up circuit is automatically disconnected from the system.

When you are troubleshooting a power supply like this, begin with the start-up circuit, which is easily identified on the schematic. An oscillator signal can be injected into the start-up circuit. This is the type of circuit that is best serviced by using a manufacturer's diagnostic chart.

Current Regulation

Instead of controlling voltage, it is also possible to use a feedback circuit to control current. An example of a current regulator is shown in Fig. 7-11. The voltage drop across the series resistor controls the bias of the transistor. That, in turn, controls the gain of the transistor and the amount of current flowing through the load resistance. An increase in current through the load resistor will cause an increase in bias and a decrease in gain. The resistance of the transistor increases with the decrease in gain, and the current through R_L is decreased to the desired value.

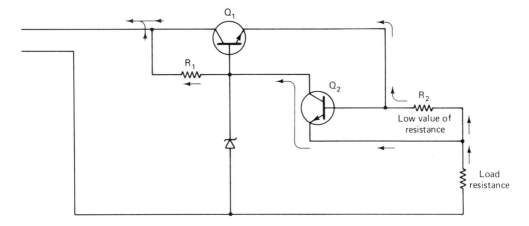

Figure 7-11   Principle of a current regulator.

FEEDBACK CIRCUITS THAT CONTROL OR ESTABLISH FREQUENCY

Oscillators are circuits that employ regenerative (positive) feedback. In other words, the feedback signal is in phase with the input signal.

Figure 7-12 shows a block diagram of a typical sine wave oscillator. Note that an amplifier is an important part of this circuit. For that reason you can often determine whether an oscillator is working simply by treating it as an amplifier. For example, start by measuring the emitter-to-base voltage and collector voltage.

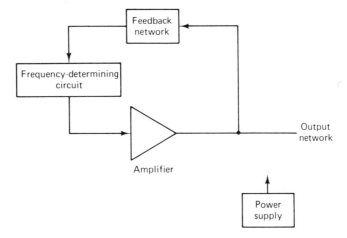

Figure 7-12   Block diagram of a sine wave oscillator.

You cannot limit oscillator troubleshooting to this approach, however. The oscillator can be working, but it can be working at the wrong frequency.

A much better approach is to use a frequency counter if you have one available. That tells you if the oscillator is working and also if it is working at the right frequency. A frequency counter is a very good investment for today's technician.

Your knowledge of electronics is one of your best troubleshooting aids, and it is a good idea to have the block diagram of the oscillator clearly in mind. If there has been a catastrophic failure in the oscillator, the amplifier may still appear to be working properly. With vacuum tube and some FET oscillators, lack of oscillation can lead to complete loss of bias. Since it is a closed-loop circuit, it is not possible to tell whether the problem is in the amplifier or in the feedback system or other supporting systems. The indication of failure is that the frequency counter shows no output frequency.

An in situ tester can be very important here. It will show that the bipolar transistor is OK and capable of amplifying and supporting oscillation.

Dip meters can be used to inject a signal into the oscillator-tuned circuit. If oscillation still doesn't occur, then it is a good likelihood that there is a problem in the feedback circuit. In its simplest form, the feedback circuit uses a transformer or autotransformer. However, the feedback in other oscillators can also be a capacitive feedback circuit. Figure 7-13 shows an example.

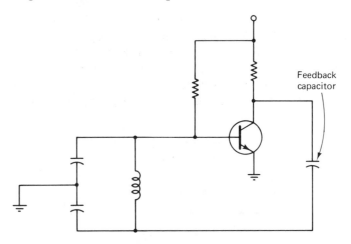

Figure 7-13   Capacitive feedback in a sine wave oscillator.

One good way to check oscillators is to open the loop and inject the required feedback signal into the amplifier. Then with the frequency meter you can check to see if that signal is going all the way around the loop and back to the point where you opened it.

The tuned circuit is usually an LC circuit for sine wave oscillators, and it is an R_C or R_L circuit for relaxation oscillators. An exception to this is the phase shift oscillator. Figure 7-14 shows an example. The feedback network for this sine wave oscillator is an R-C network.

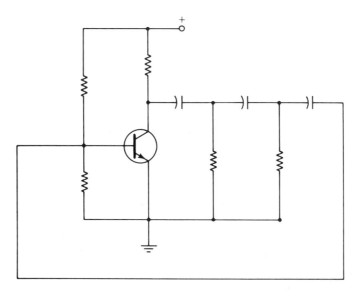

Figure 7-14   Another example of a sine wave oscillator using capacitive feedback.

Failure of the tuned circuit is not a common fault with oscillators. So, in a statistical approach, the amplifying device would be the first to suspect. This is one of the few feedback circuits where you cannot use a DC voltage to defeat the loop for troubleshooting.

In order to determine whether the oscillator is at fault for the failure of a *system*, you can disable the oscillator (usually by disabling the frequency-determining network). Then inject the required oscillator signal into the system at the output of the oscillator stage. If the system works properly, then you have oscillator troubles and your next step is to troubleshoot the oscillator circuit.

Automatic Frequency Controls (AFC)

Figure 7-15 shows a block diagram of a typical AFC system. Its purpose is to lock the oscillator frequency to a specific frequency.

At some point in the system, the frequency is sampled. To give a specific example, in a television receiver the i-f frequency (which should be about 41 megahertz) is sampled for the AFC action.

The sampled frequency is fed to an FM detector, such as a discriminator or ratio detector. The output of the FM detector is a DC voltage that is a direct function of the input frequency.

You usually see FM detectors used to demodulate an FM signal where the input frequency is varying according to an audio rate. In that type of circuit the output is an audio signal. However, in the AFC circuit the input is an r-f frequency and the output is a DC voltage.

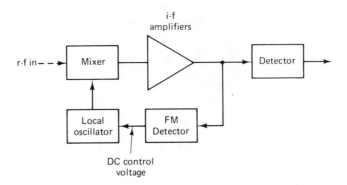

Figure 7-15   Block diagram of a typical AFC system.

Assume that the input frequency is off the desired center frequency by a certain amount. That produces a DC output. The detector DC output is used to·control a V_{CO} frequency.

To troubleshoot this system, substitute the DC from the discriminator ratio detector with the proper DC voltage. Vary the DC voltage and see if the oscillator frequency varies. (Use your frequency counter here.) If the frequency does not vary, your trouble is in the oscillator circuit. If it does vary, your trouble is in the feedback loop.

The Phase-Locked Loop

Putting the phase-locked loop in an integrated circuit opened the way to a wide variety of applications for this circuit in all fields of electronics. Rather than try to cover examples in all of the fields, two specific examples will be given.

Figure 7-16 shows a phase-locked loop used as a motor speed control. In one part of that illustration, a crystal-controlled frequency is used as the reference frequency. It is counted down before being delivered to the phase comparator.

The second input to the phase comparator comes from the feedback loop by way of the low-pass filter, amplifier, voltage-controlled oscillator, and frequency divider.

To start the discussion, it will be assumed that the input frequency from the reference, after countdown, is 10 Hz and the frequency divider is 1—that is, the V_{CO} frequency will be divided by 1.

If the two input frequencies to the phase comparator are the same, there will be a DC output. It will be passed by the low-pass filter, then amplified (in some cases), and delivered to the voltage-controlled oscillator. The DC voltage into the V_{CO} indicates that there is no frequency error. Therefore, the output of the V_{CO} will be equal to the same frequency as the reference input.

If the V_{CO} frequency is too high, there will be a DC voltage out from the phase comparator that is fed back through the loop. It will bring the oscillator frequency to the required value. Conversely, if the oscillator frequency is too low, the feedback loop will produce a DC voltage to the V_{CO} that will raise its frequency. That will make it equal to the reference input to the phase comparator.

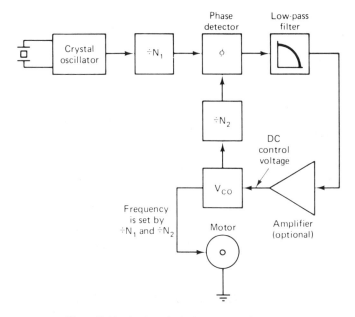

Figure 7-16 A phase-locked motor speed control.

So far, nothing really has been accomplished because the output frequency of the V_{CO} is exactly equal to the reference frequency. Since the reference is crystal-controlled there is little point in the phase-locked loop. But suppose it is desired to change the output frequency and still have it locked to a reference signal.

In this case, the reference input will still be 10 Hz. The programmable divider will now be set to divide by 10. Now, in order for the two inputs to the phase comparator to be equal, it is necessary for the V_{CO} to be operating at ten times the reference frequency. That way, when it is divided by 10, it will be equal to the reference frequency.

So the output of the V_{CO} will now be 100 Hz.

You can set the programmable divider to any division and change the oscillator V_{CO} output to a desired value you want.

If the frequency divider in the reference circuit is also changeable, it is possible to produce any desired frequency (within the range of the system) that is crystal-controlled. This is called *frequency synthesizing.*

The motor speed control requires a motor that has a speed determined by the frequency of its input signal. One example is a stepping motor in which the number of pulses per second delivered to the stepping motor determine its rpm.

Frequency synthesizers of the type just discussed are used extensively in communications equipment and in consumer electronic equipment. As an example, the tuner in a broadcast radio can be controlled by a microprocessor that controls the programmable divider. In that way the microprocessor can set the desired local oscillator frequency for any specific broadcast frequency chosen.

Troubleshooting the phase-locked loop involves opening the loop. A logical place to do that is at the input of the V_{CO}. Replace the amplifier output, if it exists, or the output of the low-pass filter.

NOTE

The amplifier is not always used. Its purpose is to make the feedback circuit more sensitive to very small changes in V_{CO} signal. However, if the circuit is too sensitive, sometimes oscillations may occur. In that case, the amplifier is not used.

A variable DC voltage supplied to the V_{CO} should make it possible to determine if the loop is working. Change the DC voltage and see if the DC output (now disconnected) changes. If not, it is likely that the oscillator isn't changing. Check it with a frequency counter as you change the DC voltage.

Make sure that the frequency reference is correct using the frequency counter. If that doesn't show the problem, you can suspect the phase comparator.

REVIEW OF SOME BASIC CIRCUITS USED IN CLOSED LOOPS

There are some basic circuits that are often used in frequency control feedback systems. A few of these will be reviewed here.

V_{CO}   Voltage-controlled oscillators have an output frequency that depends on the DC voltage delivered to their control electrode.

Discriminator/Ratio Detector.   Both are circuits that have an output voltage that is dependent on their input signal control frequency. If the input r-f signal frequency varies according to an audio signal, as in FM, the output of these circuits will be the audio modulating frequency.

Remember, however, if the input r-f frequency is unvarying, but is different from the center frequency of the discriminator/ratio detector, the output will be a DC voltage.

Phase Comparator.   This circuit provides a DC output voltage that is directly related to the difference in phase between two input signals. It is sometimes combined with automatic frequency circuitry to make a circuit called AFPC (automatic frequency and phase control).

Programmable Counter/Divider.   This is a digital circuit that performs either of two functions. It can count up or down. Also, it can be used to divide an input signal frequency by an amount determined by the control input.

Synthesizing.   This term means to make a new frequency by combining input frequencies.

SUMMARY

There are two major categories of closed-loop systems: those that control voltage and those that control frequency. When discussing closed-loop *control* circuits, it is implied that negative feedback is being used. In other words, the feedback is used in some way to reduce the output of the system.

Positive feedback is also used in systems and circuits. The best known use of positive feedback is in oscillators.

Returning to negative feedback control, both kinds (voltage and frequency) utilize a DC voltage for their operation.

Troubleshooting the closed-loop control circuits requires that the loop be opened at some point where there is a DC control voltage. A power pack or power supply is used to substitute for the DC control. Then the circuit can be treated as an open loop.

Instead of opening the loop, it is sometimes possible to clamp the DC control voltage with a fixed DC source.

Troubleshooting positive feedback oscillators is usually accomplished by taking measurements on the amplifier. A substitute oscillator frequency can be injected to determine if the frequency-sensitive networks are OK.

PROGRAMMED SECTION

Instructions for this programmed section are given in Chap. 1.

1. According to this chapter, two circuits that are especially difficult to troubleshoot are

   A. reference and sense. Go to block 7.

   B. closed-loop and intermittent. Go to block 14.

2. You have selected the wrong answer for the question in block 4. Read the question again. Then go to the block with the correct answer.

3. The correct answer to the question in block 16 is A. This type of control is called PWM (pulse width modulation).

   Here is your next question:

   A high frequency is used in switching regulator power supplies because

   A. they are easier to troubleshoot. Go to block 20.

   B. they are easier to filter. Go to block 9.

4. The correct answer to the question in block 14 is B. The closed-loop regulator is a feedback circuit used in regulated voltage power supplies. A simple example is shown in Fig. 7-7.

   Here is your next question:

   Where does the AVC/AGC DC control signal come from?

   A. detector. Go to block 12.

   B. power supply. Go to block 2.

5. The correct answer to the question in block 13 is B. Maximum output sound almost always corresponds to poor sound quality.

   Here is your next question:

   Why is a battery pack best for replacing an AVC/AGC voltage when troubleshooting a receiver?

   A. no ripple. Go to block 16.

   B. better connections. Go to block 21.

6. You have selected the wrong answer for the question in block 9. Read the question again. Then go to the block with the correct answer.

7. You have selected the wrong answer for the question in block 1. Read the question again. Then go to the block with the correct answer.

8. You have selected the wrong answer for the question in block 18. Read the question again. Then go to the block with the correct answer.

9. The correct answer to the question in block 3 is B. Components with lower values of inductance and capacitance work well to filter the high-frequency ripple in the regulator.

   Here is your next question:

   You might find a start-up circuit in a regulated power supply that has

   A. a switching regulator. Go to block 23.

   B. an analog regulator. Go to block 6.

---

10. You have selected the wrong answer for the question in block 16. Read the question again. Then go to the block with the correct answer.

---

11. The correct answer to the question in block 12 is A. The feedback voltage in an oscillator circuit is an oscillator-frequency signal.

   Here is your next question:

   The purpose of the AVC/AGC signal is to

   A. keep the gain of the r-f and i-f amplifiers constant. Go to block 22.

   B. keep the output of the system constant. Go to block 18.

---

12. The correct answer to the question in block 4 is A. The output from the detector is an AC signal sitting on top of a DC voltage. An AVC/AGC filter is used to smooth out the signal fluctuations so that the circuit output is a DC control voltage.

   Here is your next question:

   Which type of feedback circuit is not serviced by opening the loop and inserting a controlled DC voltage?

   A. an oscillator. Go to block 11.

   B. an AFC circuit. Go to block 15.

---

13. The correct answer to the question in block 18 is C. Unless you have a test procedure from the manufacturer, you have to analyze each system to determine the polarity of the AVC/AGC voltage. In some systems it is also possible to have both positive-going and negative-going AGC voltages.

   Here is your next question:

   When aligning a radio, the i-f transformers should be adjusted

   A. to give the maximum output sound. Go to block 17.

   B. to give the best fidelity of sound. Go to block 5.

---

14. The correct answer to the question in block 1 is B. Troubleshooting problems in closed-loop circuits have been discussed in this chapter.

   Here is your next question:

   Which of the following is a closed-loop voltage control system?

   A. AFC. Go to block 19.

   B. closed-loop regulator. Go to block 4.

---

15. You have selected the wrong answer for the question in block 12. Read the question again. Then go to the block with the correct answer.

---

16. The correct answer to the question in block 5 is A. A battery bias pack is one of those pieces of test equipment you have to make for yourself; they are not readily available as test equipment. Use rechargeable batteries.

   Here is your next question:
Switching regulators control output voltage (and power) by controlling

   A. the width of a pulse. Go to block 3.

   B. the amplitude of a pulse. Go to block 10.

---

17. You have selected the wrong answer for the question in block 13. Read the question again. Then go to the block with the correct answer.

---

18. The correct answer to the question in block 11 is B. The AVC/AGC voltage *controls the gain* of the amplifiers in such a way that the output is constant. As an example, the sound volume from a radio's loudspeaker is held constant even though the strength of the incoming signal varies. The constant volume is achieved by controlling the gain of the amplifiers at the signal input side. If the signal strength increases, the gain is reduced. Conversely, if the signal becomes weaker, the gain of the amplifiers is increased.

   Here is your next question:
Tuning a radio from a point where there is no station to a point where there is a station should cause the AVC/AGC voltage to

   A. become more positive. Go to block 8.

   B. become more negative. Go to block 24.

   C. cannot answer; it depends on the circuit. Go to block 13.

---

19. You have selected the wrong answer for the question in block 14. Read the question again. Then go to the block with the correct answer.

---

20. You have selected the wrong answer for the question in block 3. Read the question again. Then go to the block with the correct answer.

---

21. You have selected the wrong answer for the question in block 5. Read the question again. Then go to the block with the correct answer.

---

22. You have selected the wrong answer for the question in block 11. Read the question again. Then go to the block with the correct answer.

---

23. The correct answer to the question in block 9 is A. Start-up circuits are needed when the power for the oscillator comes from the output of the supply.

   Here is your next question:
Statistical chance of a component going bad is an important consideration when

replacing a part using the shotgun method. Another consideration is ____. Go to block 25.

24. You have selected the wrong answer for the question in block 18. Read the question again. Then go to the block with the correct answer.

25. The correct answer to the question in block 23 is how difficult it is to remove the part. Start with parts that are likely to go bad *and* are easy to replace.

You have now completed the programmed section.

TEST YOUR KNOWLEDGE

1. Name two types of feedback used in electronic systems. ____, ____
2. Name two types of circuits that use degenerative feedback. ____, ____
3. What kind of voltage is used for control in a degenerative feedback circuit? ____
4. What is the standard procedure for troubleshooting a closed-loop control system? ____
5. Name two kinds of closed-loop voltage regulator circuits for power supplies. ____, ____
6. What is an advantage of a switching regulator over the analog type? ____
7. What circuit stabilizes output volume from a speaker despite changes in input signal strength? ____
8. In reference to the circuit described in question 7, what is the DC control voltage used for? ____
9. Is the following statement correct? In an analog power supply regulator, the sense circuit is connected across the load resistance. ____
10. What type of component is used to get a reference voltage in a regulated power supply? ____

ANSWERS
TO TEST YOUR KNOWLEDGE

1. negative, also called degenerative; positive, also called regenerative
2. circuits for controlling voltage, circuits for controlling frequency
3. DC
4. open the loop
5. analog, or linear; digital, or switching
6. greater efficiency
7. AVC/AGC
8. control gain of the r-f and i-f amplifiers
9. The statement is correct.
10. zener diode

8

Hunting for the Causes
of Noise and Intermittents

CHAPTER OVERVIEW

Although noise and intermittents are two separate types of problems, they are included in this one chapter because some of the methods used to track down the two problems are the same. Furthermore, intermittents can cause noise, so they are also related in that way.

As mentioned in the preface, it is very difficult to categorize some of the test procedures used by technicians. In some cases, the categories are strictly arbitrary. Part of this chapter is devoted to identifying types of noise and types of intermittents.

Noise is not always undesirable. As discussed in Chap. 5, you can use noise as an injected signal. It is especially useful for aligning broad-band circuits.

Objectives

Some of the topics discussed in this chapter are listed here:

- How is noise related to the strength of an incoming signal?
- Where is the noise generated in a receiver?
- What causes snow and hissing noise in receiving systems?
- What do the names *Boltzmann's constant* and *Avogadro's number* mean to an understanding of noise?
- How is noise related to the bandwidth of an amplifier?
- What is $1/f$ noise?

TYPES OF NOISES AND THEIR CAUSES

It is important to understand that *most* of the noise created internally in a system is created in the front end. For example, in a television receiver, almost all of the troublesome noise generated by the circuitry is created in the television tuner.

In a video system, such as a television receiver, the noise appears as snow on the screen. In a system designed to communicate audio, the noise shows up as a hissing sound in the speaker. In other systems, such as digital and microprocessor, the internally generated noise is seen as grass on an oscilloscope display of the signal.

Noise that is created internally by circuitry is not to be confused with the noise that is created outside the system. Using television as an example, atmospheric interference creates noise that is generated outside the receiver.

The amount of snow due to internally generated noise is indirectly related to the strength of the incoming signal. A strong signal can override noise voltages. If the signal-to-noise ratio is high enough, the noise problem is eliminated.

If the signal is weak, the AVC/AGC system will raise the gain of the receiver. That also raises the output noise. This is the reason a good antenna system is needed in weak-signal areas. Hunting down noise sources and eliminating them may involve going outside the system to improve the strength of the signal being delivered to the system.

Resistor Noise

This discussion begins by considering a perfect amplifier, as shown in Fig. 8-1. This perfect amplifier increases the amplitude of an input signal, but it does not add any noise to the signal output.

The output power, measured by the wattmeter, will show no power output when there is no input signal. This is to be contrasted with an actual amplifier. With a real amplifier, the wattmeter would show some output power without an input signal. That output power would be due to noise generated within the amplifier. But, remember, we are talking about a perfect amplifier in this case. The concept of the perfect ampli-

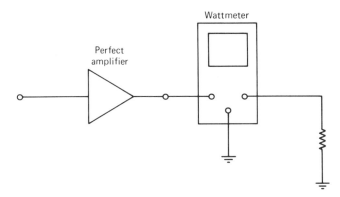

Figure 8-1 The idea of a perfect amplifier.

fier is actually used to define noise measurements. So it is not just a hypothetical case that has no real meaning.

In Fig. 8-2 a resistor has been connected across the input of the amplifier. Now the output wattmeter shows a measurement, *even though the amplifier still does not produce any noise.* The output measurement is called *noise power.* It is the result of the amplifier increasing the amplitude of the noise generated by the resistor.

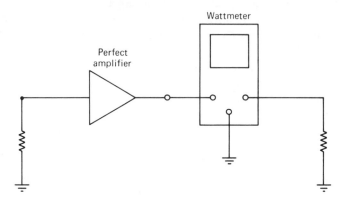

Figure 8-2   Connection of the resistor across the input terminals causes noise to be injected.

There is no current flowing in resistor (R) in the circuit. Obviously, then, the noise is not created by electrons "bumping into atoms and other electrons" as they move in a current flow. Where, then, does the noise come from?

To understand this, consider the model of a resistor shown in Fig. 8-3. There is no voltage applied across this resistor by an outside source of power. At room temperature (and, in fact, at *all* temperatures) the atoms in the resistor are in continual motion, called *Brownian motion.* At room temperature some electrons escape from the atoms for a short period of time. They flow through the material until they are captured by another atom that has lost an electron. This motion of electrons for a very short duration of time is called *intrinsic current.*

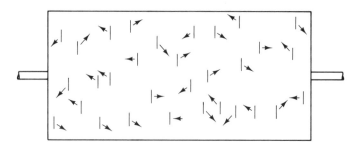

Figure 8-3   A model of a resistor at room temperature.

The motion of the electrons is random. However, at any specific instant of time there will be more electrons moving in one direction than in the other. At that instant of time a voltage is created across the resistor.

In the next instant there may be a majority of electrons flowing in the opposite direction. That reverses the polarity of the voltage across the resistor.

Remember, these electrons are flowing because they have obtained enough energy to escape from their moving atoms.

If you increase the temperature of the resistor there will be more electrons moving in the intrinsic current. Whenever the net amount of intrinsic current flowing in one direction is greater than that flowing in the other direction, there is an output voltage. The higher the temperature the higher the amplitude of the voltage. That is because at a higher temperature there is more intrinsic current.

Over a period of time the voltages created by the random electron flow in the resistor results in a random fluctuation of voltage at the input of the amplifier in Fig. 8-2. That random fluctuation is amplified by the (perfect) noiseless amplifier and creates the output noise power.

There is an immediate practical application to the noise that has just been discussed. Certainly, most amplifiers have resistors at their input terminal and these resistors will produce noise. Resistors are something to consider when tracking amplifier noise. That noise can be very troublesome in some systems. If the noise is greater in amplitude than the incoming signal, then the signal-to-noise ratio is not satisfactory.

It should be obvious that you must be very careful when you replace resistors in certain circuits. For example, what happens if you replace a film resistor at the input of an amplifier with a carbon composition resistor (which creates more random noise voltage)? The answer is that you will likely introduce enough additional noise to change the signal-to-noise characteristic of the system.

Good technicians are very careful to use exact replacements of resistors, especially for those parts of the system near a signal source.

This discussion could be applied to the input circuit of any industrial analog system or digital system. In that case, the input might be from a transducer. The system is used to amplify the relatively weak transducer input to a higher and more useful amplitude.

As with the case of the television receiver, the noise created by resistance across the input terminals of any amplifier can render it useless. This is especially true if a careless technician replaces a resistor with the wrong type.

Here is another practical application of what has just been discussed: A well-designed communications receiver has just been installed in a new location and the technician connects the antenna terminals to an outside antenna. Immediately *the technician has injected noise into the system because of the antenna resistance.* Some types of antennas introduce more noise than others, and the technician should make sure to determine which antennas produce the highest possible signal-to-noise ratio in the system.

Remember this important point: Any resistor, conductor, or semiconductor device connected across an input of a small signal amplifier (that is, an amplifier that has high voltage gain) will introduce noise into that amplifier. *When looking for the source of noise, you should start by going to the source of the signal.*

Of course, if an amplifier has been working for some time and suddenly it begins to produce output noise, the noise is not necessarily caused by a change in the input resistance. However, that is always a possibility that should be considered. Another likely cause is that an amplifier gain has been reduced. That, in turn, can change the signal-to-noise ratio to an undesirable value.

What Is Boltzmann's Constant?

Whenever you see an equation for noise in an electronic system, you are sure to find a *k*—Boltzmann's constant. Technicians don't like to work with things they don't understand, so a brief discussion of Boltzmann's constant will be given here.

Amedeo Avogadro (1776–1856) was an Italian physicist who startled the world of science with the following declaration: *Equal volumes of gases at equal temperatures and equal pressures have exactly the same number of gas molecules.*

That wasn't much more than a guess when he first proposed it, but it turned out to be a very good guess. It has been verified a number of different ways by scientists. By defining the volume and pressure and temperature, the number of molecules was established to be 6.02×10^{23}. This is called Avogadro's number.

Avogadro's guess is now called Avogadro's law. It was not immediately accepted but eventually became a very important part of modern theory.

There is another very important law that can be stated as a simple equation:

$$pv = RT$$

If you multiply the pressure of a gas (*p*) by its volume (*v*), you get the temperature of the gas (*T*) (measured in degrees Kelvin)* multiplied by a constant *R*. The value of *R* is the same for all gases, and it is called the gas constant.

You can determine the value of *R* by multiplying the pressure and volume of any gas and then dividing the product by the temperature of the gas. No matter which gas is used, you always get the same value.

Two very important constants have been defined: *the gas constant (R) and Avogadro's number (N)*. Neither is equal to zero, so you can divide one by the other and get a third constant (*k*):

$$k = \frac{R}{N}$$

That value (*k*) is Boltzmann's constant. It is used in many scientific equations.

You will often see Boltzmann's constant as a multiplier for absolute temperature (*T*). That makes *kT* a *unit of energy*. It is used, among other things, in discussions of energy levels of atoms in materials.

---

*Degrees Kelvin is the number of degrees above absolute zero temperature.

Degrees Kelvin = 273° + Degrees C

The amount of noise generated by a resistor increases with temperature. That's because the atoms have more energy at higher temperatures. The energy levels are expressed as kT. So why does kT show up in equations for noise in transistors, resistors, and so on? Because the amount of noise generated in those devices is directly related to the energy of the atoms (and electrons).

Using Boltzmann's constant, scientists are able to find out the expected electron energy and, hence, the amount of noise that will be created in various components. It is interesting to look at the equation for the amount of noise generated by a resistor.

This equation is not given so that you can, armed with a calculator, determine the amount of noise voltage in various parts of the equipment. As in many cases, an equation serves only one purpose: to show you the relationship between some of the variables in the system. Here is the equation:

$$\text{Noise power} = kTB \text{ watts}$$

In this equation, k is Boltzmann's constant, T is absolute temperature (in degrees Kelvin), and B is the bandwidth (in hertz).

The equation for noise voltage generated in a resistor is

$$\text{Noise voltage} = \sqrt{4\,kTBR}$$

where kT and B have the same meaning as in the noise power equation, and R is the resistance of the resistor.

Various names have been given to this noise generated by a resistor at room temperature—*Johnson's noise, thermal noise, thermal agitation noise,* or *white noise.*

The term *white noise* is a misnomer. It comes from the idea that the noise signal produced by the random motion of electrons has a random frequency and amplitude. It is supposedly the same random relationship as the color white, which is supposed to contain all in the colors of the rainbow spectrum.

However, white is not white—at least as it is perceived by a human being. You could not make something look white on television by putting equal parts of red, green, and blue.

But the name *white noise* persists. Think of it as a very wide spectrum of frequencies and amplitudes created by the random motion of electrons.

The equations just given clearly show the effect of temperature on the noise created in a system.

An Example of Noise

To get a direct idea of the effect of connecting a resistor across amplifier terminals, consider this example. A 5-megohm resistor connected across an amplifier that has a bandwidth of 5 megahertz will inject about 10 microvolts of noise into the amplifier at room temperature.

Ten microvolts can be an appreciable part of the total input of an r-f amplifier. In some systems the total amount of useful input r-f signal is about 50 μV delivered to the antenna terminals. So a noise signal of 10 μV would account for about 20 percent of the total input to the antenna terminals.

The equation also shows that *the amount of noise can be increased by increasing the bandwidth of the amplifier.* As you would expect, increasing the resistance of the resistor will also increase the amount of noise injected.

Johnson's noise is not the only noise you have to worry about in an amplifier. Refer to Fig. 8-4, which shows the types of noise and their distribution in a typical

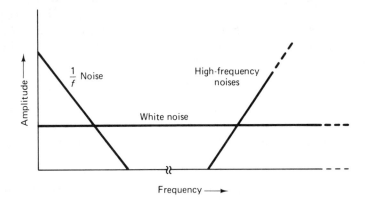

Figure 8-4   Three kinds of noise.

electronic amplifier. Note that white noise has a constant amplitude throughout the bandwidth.

When current flows through a resistor, very small instant-to-instant changes in resistance cause a varying noise voltage. This is a source of $1/f$ noise.

If you heat the cathode of a vacuum tube, electrons are emitted from its surface. Unfortunately, they don't all leave like soldiers marching in unison. At any instant of time the number leaving the cathode will be different from the number at any other instant. This produces a special kind of noise referred to as *flicker noise,* or $1/f$ noise.

You might think you get away from flicker noise with transistors and field effect transistors.

The cause of this type of noise in semiconductor devices is thought to be the flow of charge carriers across the surface of (rather than through) the device. It is still a form of noise.

Flicker noise is present in all amplifying devices. Unfortunately, there isn't much you can do about it, but it does decrease in amplitude as the frequency increases. Flicker noise is primarily a problem at the low-frequency end of the spectrum.

Another problem with amplifying devices is called *partition noise.* It can best be described by using the model of a bipolar transistor. See Fig. 8-5. Note that of all the electrons leaving the emitter area a small part go to the base and the rest go to the

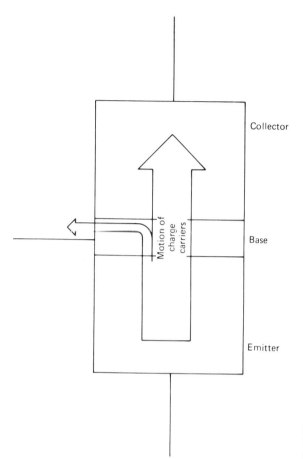

Collector

Motion of charge carriers

Base

Emitter

Figure 8-5 The cause of partition noise is illustrated in this bipolar transistor model.

collector region. The problem is that from instant to instant the number that go to the base vs. the number that go to the collector changes.

That means that the base current is changing from moment to moment by a very small amount. The variations in current are a source of transistor noise. The small base current changes are amplified due to the transistor beta. So the amount of amplified output noise can be an important factor in the total amount of amplifier noise.

Partition noise is also a problem in vacuum tubes. In Fig. 8-6 you see electrons sticking to the control grid of a triode vacuum tube and returning to the cathode through the grid resistor. In a manner similar to the biopolar transistor, the number of electrons that actually flow through the grid circuit is subtracted from the output (plate) current. So there is a moment-to-moment change in current through the grid resistor.

Moment-to-moment variations in the partition current flowing through R cause

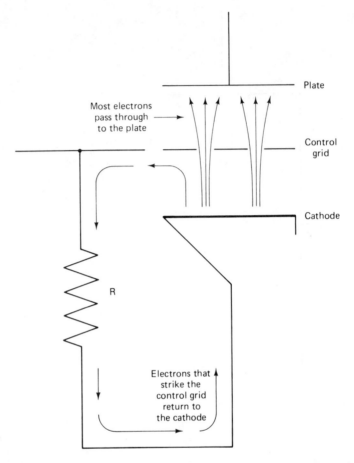

Most electrons
pass through ———→
to the plate

Plate

Control
grid

Cathode

R

Electrons that
strike the
control grid
return to
the cathode

Figure 8-6 Partition noise is also a problem in vacuum tubes. In this illustration
some electrons return to the cathode through the grid resistor. The number varies
from moment to moment.

a noise voltage to appear across that resistor. The noise voltage is amplified by the
tube.

In a field effect transistor there is no motion of electrons into the gate region.
That means that there is no partition noise in those devices. The same is true of MOS-
FETs. That is the reason why you see field effect transistors used extensively in modern communications and other high-frequency electronic systems, especially in r-f
circuits.

There are other kinds of noise generated in electronic systems, but the most important kinds have been discussed as far as the noise generated within the system is
concerned.

External Noise

Noise generated outside the system is very often some form of impulse noise. Examples are static created by lightning and noise generated by machinery. Impulse noise, as with any other pulse, contains a very large number of harmonic frequencies. It is a great disturbance in electronic systems.

Not much can be done about impulse noise for a given system. Some intelligent use of antenna theory may help. Coaxial cable prevents the impulses from being picked up by the transmission line. The shield prevents the line from acting like an antenna.

Impulse noise is usually horizontally polarized, so long horizontal runs should not be used in a receiver in an antenna system.

An Example of a Noise Problem

Now consider a system that was working OK at one time and all of a sudden requires servicing because of internal noise. This may be due to a transistor that has had a beta change that reduces the signal-to-noise ratio. It could also be due to a bad component, such as a resistor, which is becoming too hot.

The question is: How do you find the noise source? It is assumed that all preliminary tests have been done and the power supply voltages measured.

Start troubleshooting noise problems by defeating all closed-loop circuits in the system. If, for example, the system has an automatic gain control, it is important to defeat that circuit.

Automatic frequency controls can also carry noise signals in a feedback loop because of the wide bandwidth of the noise. So it is necessary to defeat the automatic frequency control as well as automatic gain control. The best way to do this is to replace the DC voltage in these feedback systems with a DC supply voltage—preferably from a battery pack that will not add hum or noise.

Receivers are not the only electronic systems that employ negative feedback. You will find negative feedback in industrial speed controls, hearing aids, public address systems, and a wide variety of other systems. In all cases these feedback circuits must be located and defeated when looking for the source of noise.

Having defeated all of the closed-loop systems, you can begin to look for the noise source by *starting at the input of the system.* For this type of problem, the idea of starting in the middle of the system should *not* be used. Remember that noise is created at the input end of the system, so there is no use in wasting time in the middle.

In an industrial system, start at the transducers; in a communications system, start at the antenna terminals; in an audio system, start at the microphone. Again, as shown by these examples, you start at the input.

Moving away from the input, you disable the circuits one at a time. If the noise stops, then the noise source is in front of the circuit you just disabled. (Disabling a circuit usually means to ground the signal input or, in the case of bipolar transistors, to short the emitter to the base.)

As you move through the system, listen to the output noise. This is best done by turning the system gain to the maximum value, and at the same time, defeating the input signal from transducers, antennas, or other signal sources.

When you move past the offending circuit, the amount of noise in the output should be much lower.

Another method of finding noise is to think of the undesired noise as being an injected signal. Use the signal tracing method to find the source. (In this method you can start in the middle of the system and work toward the input). When you pass the source of the noise (going toward the front end), you will no longer be able to detect it. That tells you where it is located.

INTERMITTENTS

Intermittents can be very frustrating troubleshooting problems in electronics. Think of the intermittent as being a switch that turns a signal on and off in the system. Your job is to locate the switch.

The problem can be a break in the printed circuit board, a cold solder joint, a defective tube, transistor, or FET, a broken wire, or an intermittently shorted component. In fact, the variety of sources of intermittents is almost endless. An important thing to remember is that intermittents, like noise sources, will be very difficult to find if you do not defeat the closed-loop feedback circuits within the system. That is the first thing to do after performing preliminary checks and measuring the supply voltage.

Keep in mind that the power supply can be the source of the intermittent. So it is a good idea to disconnect the power supply and look at its output on an oscilloscope to see if it is constant.

Starting at the input, as in the case of noise, you can look and listen for the intermittent as you move through the system. While you are doing this, it is a good idea to flex printed circuit boards and move components with an insulated stick to try to stimulate the intermittent.

Intermittents can be temperature related, so technicians often use a heat source (such as a hand-held hair dryer) to heat different sections of a circuit as a method of inducing or stimulating the intermittent action. This is especially useful in a type of intermittent that will only occur at higher temperatures. For example, when the system is in its cabinet, it will display intermittent behavior, but when it is out on the workbench, the cooler ambient temperature prevents the intermittents from occurring.

Technicians have to make their own decision about spraying supercoolants on components to stop noise or intermittent interference. They cool the component with a freon spray. That stops the intermittent if it is the type that occurs with a high temperature.

If you decide to use these supercoolants, remember this: *They can be very dan-*

gerous because they can produce instant frostbite! Furthermore, if inhaled into the lungs they can produce permanent lung damage!

Many technicians avoid them. Others are prepared to exercise the safety measures necessary for their use.

SUMMARY

Much of the noise in a system is generated by resistors, diodes, transistors, and other semiconductor devices. The amount of noise depends on some very basic theoretical considerations.

Avagadro's law and Boltzmann's constant were originally related to the laws of gases. However, it is now possible to use these laws to calculate the amount of noise to be expected from a resistor or other semiconductor device.

A perfect amplifier does not add any noise to a system. A resistor connected across the input terminals of a perfect amplifier will result in a noise power at the amplifier output. So noise can be added by adding semiconductors at the input.

More important, replacing a semiconductor device with the wrong type can produce an increase in system noise.

Almost all of the noise in a system starts at the signal input end.

Before chasing noise or intermittents through a system, disconnect all closed-loop circuits.

PROGRAMMED SECTION

Instructions for this programmed section are given in Chap. 1.

1. Is the following statement correct? Noise signals in electronics are always undesirable.

A. The statement is correct. Go to block 7.

B. The statement is not correct. Go to block 14.

---

2. You have selected the wrong answer for the question in block 13. Read the question again. Then go to the block with the correct answer.

---

3. The correct answer to the question in block 4 is A. White noise, or Johnson's noise, occurs over a wide range of frequencies, but $1/f$ noise decreases in amplitude as the frequency increases. The name *pink noise* is sometimes used to refer to $1/f$ noise.

    Here is your next question:

Partition noise is not a problem in

A. vacuum tubes. Go to block 5.

B. bipolar transistors. Go to block 16.

C. field effect transistors. Go to block 8.

---

4. The correct answer to the question in block 9 is A. Resistors and semiconductor amplifying devices are two sources of Johnson's noise.

    Here is your next question:

Is Johnson's noise an example of $1/f$ noise?

A. No. Go to block 3.

B. Yes. Go to block 11.

---

5. You have selected the wrong answer for the question in block 3. Read the question again. Then go to the block with the correct answer.

---

6. You have selected the wrong answer for the question in block 8. Read the question again. Then go to the block with the correct answer.

---

7. You have selected the wrong answer for the question in block 1. Read the question again. Then go to the block with the correct answer.

---

8. The correct answer to the question in block 3 is C. In vacuum tubes the greater the number of electrodes the greater the noise. Pentodes are noisier than triodes.

    Bipolar transistors have partition noise.

    Field effect transistors do not have partition noise, so they are often used in the early stages of an electronic system such as a receiver.

    Here is your next question:

What is the relationship between resistor noise and bandwidth?

A. The greater the bandwidth the greater the noise. Go to block 13.

B. Resistor noise is not related to bandwidth. Go to block 6.

---

9. The correct answer to the question in block 15 is A. Noise would be injected into the r-f amplifier because of the antenna resistance.

Here is your next question:

Johnson's noise is a form of white noise. This statement is

A. correct. Go to block 4.

B. not correct. Go to block 17.

---

10. The correct answer to the question in block 13 is B. According to Avogadro's law, there will be an equal number of molecules in both chambers. This information is part of the discussion that explains the equation for noise in a semiconductor.

Here is your next question:

Flicker noise is produced by all amplifying devices. It is an example of

A. white noise. Go to block 21.

B. pink noise. Go to block 20.

---

11. You have selected the wrong answer for the question in block 4. Read the question again. Then go to the block with the correct answer.

---

12. You have selected the wrong answer for the question in block 15. Read the question again. Then go to the block with the correct answer.

---

13. The correct answer to the question in block 8 is A. Refer to the equations for noise in this chapter.

Here is your next question:

Two chambers with identical volumes are filled with gases, hydrogen in one and oxygen in the other. They have the same temperature and the same pressure. Which of the following is correct?

A. There will be more hydrogen molecules than oxygen molecules because the hydrogen atoms are smaller. Go to block 18.

B. There will be the same number of molecules in both chambers. Go to block 10.

C. There is no way of telling which chamber. Go to block 2.

---

14. The correct answer to the question in block 1 is B. Noise signals are used for aligning amplifiers and detector circuits. Some technicians use noise generators for signal injection. The noise signal is ideal for this purpose because the bandwidth is so wide that it works for all types of amplifiers and it is not necessary to switch from frequency to frequency as in the case of signal generators and function generators.

Here is your next question:

How does an AVC/AGC fault cause noise in a receiver system?

A. It injects noise into the r-f and i-f stages. Go to block 19.

B. It causes the gain of the r-f and i-f amplifiers to be so high that their internal noise overrides the signal. Go to block 15.

---

15. The correct answer to the question in block 14 is B. Actually, both answers are correct, but B is more likely to result in distracting noise. Also, a defective AVC/AGC system can shut down the receiver by turning the r-f and i-f amplifiers off with excessive bias.

    Here is your next question:

    Is it possible to inject Johnson's noise into a receiver just by connecting an antenna to the receiver input terminals?

    A. Yes. Go to block 9.

    B. No. Go to block 12.

---

16. You have selected the wrong answer for the question in block 3. Read the question again. Then go to the block with the correct answer.

---

17. You have selected the wrong answer for the question in block 9. Read the question again. Then go to the block with the correct answer.

---

18. You have selected the wrong answer for the question in block 13. Read the question again. Then go to the block with the correct answer.

---

19. You have selected the wrong answer for the question in block 14. Read the question again. Then go to the block with the correct answer.

---

20. The correct answer to the question in block 10 is B. If you know the types of noise ajd at what frequencies they occur you may be able to find its cause. Some amplifying devices are noisier than normal. The same is true of resistors and other components. Tracking down these sources of noise can be frustrating. You can reduce the time required for jobs like these by following a few basic steps.

    Here is your next question:

    An intermittent can be caused by a very small crack in a conductor on a printed circuit board. What are some things to do to locate the problem? _____ Go to block 22.

---

21. You have selected the wrong answer for the question in block 10. Read the question again. Then go to the block with the correct answer.

---

22. Here is the correct answer to the question in block 20: Flex the board lightly. Use an insulated stick to move components while observing and listening for evidence of the intermittent.

    Bridge conductors on the board and bridge components. This may not eliminate the problem, but observe carefully to determine if the problem is reduced. A heat gun may be useful.

If you can reduce the problem to a specific area, but still can't identify the specific component or conductor, you may have to resort to the shotgun method.

You have now completed the programmed section.

TEST YOUR KNOWLEDGE

1. Is flicker noise present in an FET device? ____
2. Will increasing the bandwidth of an amplifier increase its noise? ____
3. Is it possible to increase the noise in a system by using the wrong kind of replacement resistor? ____
4. Name two laws that are used to explain the amount of noise generated by a resistor. ____, ____
5. Is the following statement true? Most of the noise that is generated inside a system is introduced at the output end. ____
6. Which is noisier, a pentode tube or a triode tube? ____
7. Before tracking noise or intermittents, disconnect or defeat all ____ circuits.
8. As a general rule, which of the following is best as a DC source for defeating a closed-loop circuit? ____.
   A. rectifier supply
   B. battery pack
9. Partition noise is an example of ____.
   A. white noise.
   B. pink noise.
10. Random motion of atoms in a material is called ____.

ANSWERS
TO TEST YOUR KNOWLEDGE

1. yes; however, there is no partition noise in FET devices.
2. yes; noise is directly related to bandwidth.
3. yes
4. Avogadro's law, the gas law. Note: Boltzmann's constant is not a law, but it is an important factor in determining noise.
5. not true; it is introduced at the input end.
6. a pentode tube; the more electrodes in a tube the greater the amount of noise it generates.
7. feedback
8. B; there is less chance of hum and noise when a battery pack is used.
9. B
10. Brownian motion

9

Servicing Digital Logic and Microprocessor Equipment

CHAPTER OVERVIEW

In many ways troubleshooting digital logic and microprocessor circuits is the same as troubleshooting analog circuits. Many of the procedures already discussed will work. For example, signal injection, signal tracing, symptom analysis, and diagnostics are all useful for tracking down troubles in a digital or microprocessor circuit.

The main difference is in the types of signals that you are dealing with and the types of test equipment that are used.

There are only two levels of signal voltage to be dealt with: logic 1 and logic 0. In most of the logic systems, logic 1 is the positive power supply voltage. In today's systems that usually means +5 volts. However, there are some specialized circuits where other voltages are used.

Logic 0 is almost always zero volts.

When you first learned about linear systems, you had to learn some basic things in order to get started. For example, you had to study and learn about the following:

- Mathematics
- Power sources
- Basic components
- Amplifiers, oscillators, and other circuits
- Types of signals
- Basic test equipment
- Test procedures

To work in digital systems, you should cover the same range of basics. Your understanding of basics is the key to efficient troubleshooting in all types of systems.

In this chapter you will review the basics of digital systems, including the microprocessor. Many of the same fundamental concepts for linear systems will be followed for digital circuits.

Objectives

Some of the topics discussed in this chapter are listed here:

- What is the problem of propagation delay?
- What are the important logic gates you should know?
- What is an ENABLE?
- What do multiplexers do?
- What are flip-flops?
- What does a microprocessor do?

LOGIC SYSTEMS

It is not possible to offer a complete course in digital and microprocessor electronics in a single chapter. However, the basics covered here are essential for logic circuit troubleshooting in the same way that basics are needed for troubleshooting linear systems.

The mathematics in basic linear courses involves Ohm's law, calculation of power, and other ideas. The mathematics is important for understanding relationships between parameters rather than for making calculations.

Power Sources and Propagation Delay

Three major logic families are in use today. They are similar in regard to their use of basic components—called *gates*—but differ in other ways. For example, the power supply requirements are different for each. This is shown in Table 9-1.

There are CMOS systems that are designed especially for operation at +5V. They are used extensively in microprocessor systems.

In addition to the power supply voltage requirements, there is a difference in propagation delay for each family. *Propagation delay* is the time required for the output of a logic component to reflect a change that has occurred at the input. The ECL family is the fastest, and the CMOS family is the slowest. However, you will not be required to measure these times.

You should know about the difference in voltages and times because it explains why you can't substitute a component from one family into a system made with another family. For example, there are CMOS logic integrated circuits that have the

TABLE 9-1

Voltage	Family		
	TTL[1]	CMOS[2]	ECL[3]
Logic 0	0V	0V	0V
Logic 1	+5V regulated	4.5–12V may not be regulated	−5V

[1]TTL—Transistor-Transistor Logic.
[2]CMOS—Complementary Metal Oxide Semiconductor.
[3]ECL—Emitter-Coupled Logic.

same pinouts as corresponding TTL ICs, but you must not substitute one for the other.

The Logic Gates

Logic gates are to digital systems what amplifiers and oscillators are to analog circuits. They are the basic building blocks of all digital systems. As an example, the most complicated computer in the world is made with basic logic gates and a few specialized circuits (which are also made by logic gates).

Here are some things you must know about gates in order to be able to troubleshoot those circuits efficiently:

- You have to know the symbols so you can read the schematic drawings.
- You must know, and be able to read, the truth tables.
- You should recognize a circuit for a gate when you see one. Not all gates are made with integrated circuits.

There is no doubt that Boolean algebra is very useful for learning logic circuitry. However, it is questionable as to whether it is much help in troubleshooting. The reason for saying this is that many excellent technicians seem able to troubleshoot quite well without it. So, when Boolean expressions are included with the diagrams, think of them as being a shorthand way of expressing the logic relationships.

Figure 9-1 shows the most important characteristics of the AND, OR, and INVERTER gates. Observe that there are three kinds of symbols used in industry. Depending on the kind of system you are troubleshooting, you will need to know at least one of these symbols and possibly all of them. For example, in industrial electronics, you should know both the NEMA (National Electrical Manufacturers Association) and ANSI (American National Standards Institute) symbols, as well as the basic MIL (Military) or IEEE (Institute of Electrical and Electronics Engineering) symbol.

You must know the expected output for all of the possible combinations of inputs. You don't have time to look these up when you are troubleshooting, so they have

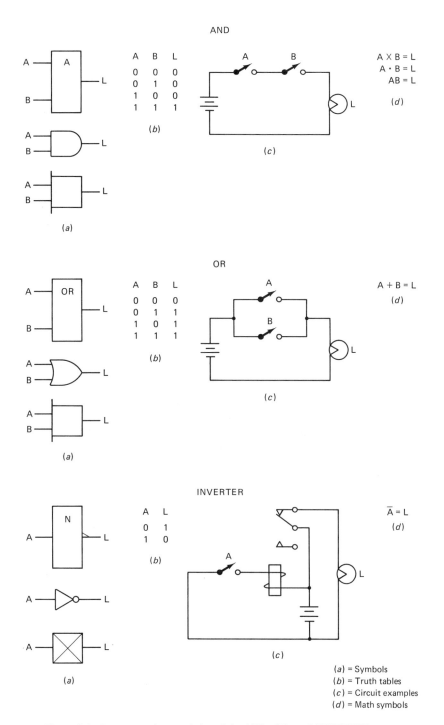

AND

A	B	L
0	0	0
0	1	0
1	0	0
1	1	1

(b)

A X B = L
A · B = L
AB = L

(d)

(c)

(a)

OR

A	B	L
0	0	0
0	1	1
1	0	1
1	1	1

(b)

A + B = L

(d)

(c)

(a)

INVERTER

A	L
0	1
1	0

(b)

$\overline{A}$ = L

(d)

(c)

(a) = Symbols
(b) = Truth tables
(c) = Circuit examples
(d) = Math symbols

(a)

Figure 9-1 Important characteristics of the AND, OR, and INVERTER gates.

to be committed to memory. Be sure you understand the special symbols that are used for inverting, as illustrated with the INVERTER gate.

An overbar is a Boolean algebra way of indicating that the signal is inverted. For example, NOT A, or $\overline{A}$, means the inverse of A. You will see this symbol used extensively in microprocessor systems. For example, $\overline{CE}$ means NOT CHIP ENABLE. This means that a logic 0 must be applied in order to enable the chip. The logic 0 is then inverted as indicated by the overbar on the symbol.

Authors are fond of calling these *active lows* and *active highs*. This simply means that an active low would involve applying a logic 0 (or connecting to common) in order to get the system into operation. That type of terminal does not always have an overbar on its designation. Figure 9-2 compares active low and active high terminals.

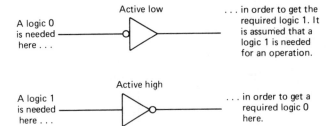

A logic 0 is needed here . . .   Active low   . . . in order to get the required logic 1. It is assumed that a logic 1 is needed for an operation.

A logic 1 is needed here . . .   Active high   . . . in order to get a required logic 0 here.

Figure 9-2   The difference between active low and active high.

Figure 9-3 shows the important things to know about NAND, NOR, and EXCLUSIVE OR gates. Be sure you have these memorized.

The EXCLUSIVE NOR gate illustrated in Fig. 9-4 is sometimes called a logic comparator. It must have identical inputs in order to get a logic 1 output.

Compare the EXCLUSIVE NOR and the EXCLUSIVE OR truth tables in Table 9-2. Note that you cannot get an output from the EXCLUSIVE OR gate if the inputs are at the same logic level. Another name for an EXCLUSIVE OR gate is HALF ADDER. The name comes from its use in computer circuits.

Logic Circuits You Should Know

In addition to the basic gates, there are some basic logic circuits you should be able to recognize. Specifically, you should know the purpose of these circuits and their input and output signals.

They are called circuits in this book because they are made with combinations of gates. However, the gate circuit constructions of the circuits are not discussed here. From the standpoint of troubleshooting, the internal makeup of these circuits is of little help in determining if they are working properly.

Three-State Devices

For three-state devices, also called tristate devices, see Fig. 9-5. These circuits have the ability to open a line that carries logic pulses.

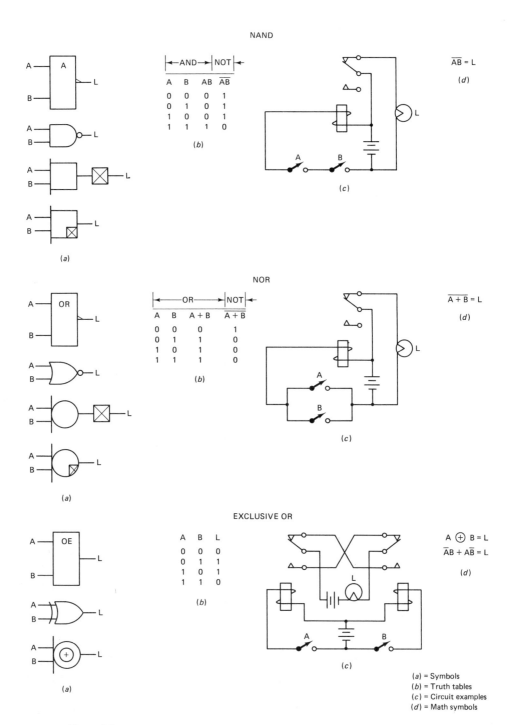

Figure 9-3 Important characteristics of NAND, NOR, and EXCLUSIVE OR gates.

(a) = Symbols
(b) = Truth tables
(c) = Circuit examples
(d) = Math symbols

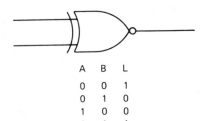

A	B	L
0	0	1
0	1	0
1	0	0
1	1	1

Figure 9-4 Symbol and truth table for a logic comparator.

TABLE 9-2

EXCLUSIVE OR

Inputs	Output
A B	L
0 0	0
0 1	1
1 0	1
1 1	0

EXCLUSIVE NOR

Inputs	Output
A B	
0 0	1
0 1	0
1 0	0
1 1	1

The circuit in Fig. 9-5(a) is called a *three-state buffer.* It passes the signal, without modification, when the disable terminal is at a logic 0 level. When a logic 1 is delivered to the disable terminal, the input signal is prevent from passing through. Also, the output terminal goes to a high impedance (open-circuit) condition.

The circuit in Fig. 9-5(b) has an active low input on the disable terminal. So it is necessary to deliver a logic 0 to the disable terminal to prevent the signal from passing through and to put the output in a high impedance state.

The circuit in Fig. 9-5(c) inverts the data when the disable terminal is at logic 0.

An example of the use of tristate logic is shown in Fig. 9-6. The data output lines from the integrated circuit pass easily when the disable terminal is at logic 0.

If data is introduced to the line by circuit X, the integrated circuit could be destroyed if a logic 0 and logic 1 appeared on the line at the same time. However, the three-state devices can be used to isolate the IC from the lines while circuit X is in operation. (In some cases it will be necessary also to have tristate devices at the output of circuit X.)

Here are some definitions of important terms that you should know.

ENABLES, like three-state devices, permit a logic signal to be either passed

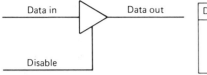

Data	Disable	Output
0	0	0
1	0	1
0	1	HI Z
1	1	HI Z

(a)

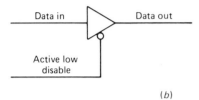

Data	Disable	Output
0	0	HI Z
1	0	HI Z
0	1	0
1	1	1

(b)

Data	Disable	Output
0	0	1
1	0	0
0	1	HI Z
1	1	HI Z

(c)

Figure 9-5 Examples of three-state devices.

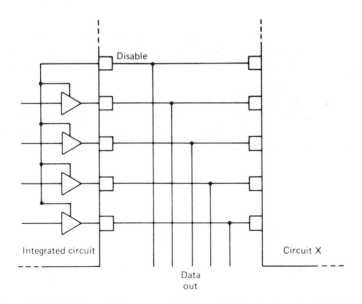

Figure 9-6 The tristate devices protect the integrated circuit when circuit X is delivering data out.

201

through or blocked. The output depends on the ENABLE input signal and the choice of AND or NAND gate.

In the circuits shown in Fig. 9-7, an open switch prevents the input signal from passing through. The output will always be logic 0 if it is made with an AND gate. It will always be 1 if it is made with a NAND gate. A closed switch enables the circuit.

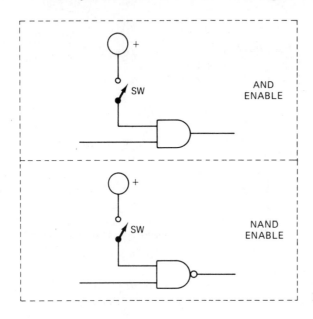

Figure 9-7 Two ENABLE circuits.

Observe that the output is inverted for the NAND compared to the AND circuit.

Registers are short memories. They are used to hold a digital number for a certain period of time. Registers are used extensively in internal microprocessor circuitry as well as in digital circuits. There are four possible registers, as indicated in Fig. 9-8.

Multiplexers are like switches. They enable a high number of inputs to be delivered to a single output. Figure 9-9 shows a mechanical interpretation of the multiplexer action. A special kind of multiplexer, called a *demultiplexer,* can be used to deliver a single input to a number of outputs.

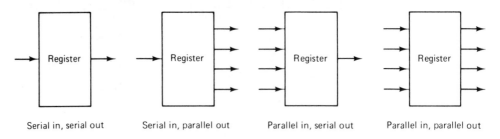

Figure 9-8 The four kinds of registers.

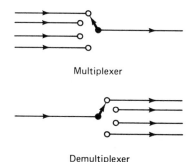

Multiplexer

Demultiplexer

Figure 9-9 Switch equivalent circuits for the multiplexer and demultiplexer.

Frequency counters and *frequency dividers* are used extensively in electronic circuits. They are usually made with flip-flops.

Frequency counters count *up* in binary numbers. *Programmable counters* can be set, using a binary code, to count to a predetermined value and then reset.

Frequency dividers count *down*. They are used to divide an input frequency to a lower value. Like counters, they use flip-flops. For an application of frequency dividers, refer to the discussion on phase-locked loops in Chap. 7.

Flip-flops are sometimes called bistable circuits. The output is at either of two levels usually referred to as *high* and *low,* or Q and $\overline{Q}$. Switching from one output to the other may involve the use of a clock signal and a dynamic circuit, or it may be done with a static combination of input signals.

If a clock signal is delivered to a *toggled flip-flop,* the output is one-half the input frequency. In counters and dividers many flip-flops may be connected internally.

Data flip-flops are used extensively in static memories (without clock signal) such as *static RAMS.*

Logic comparators are used to compare two logic signals. They produce an output only when the inputs are identical. The simplest logic comparator is the EXCLUSIVE NOR, which was discussed earlier in this chapter.

Decoders and encoders are often used in display circuits, but they also have other applications. A *decoder* takes a logic level input and converts it to a specific output. A good example is the seven-segment decoder used in a basic counting circuit, as shown in Fig. 9-10. Here the input is a binary count, and the output controls the segments in such a way that a decimal number is displayed to represent the binary input.

Encoders are the opposite of decoders. They are used to convert some input into a logic count or a logic number. A good example of encoders is in the keyboard electronics that converts a keystroke into a binary number that can be understood by a microprocessor. Each key on the keyboard is *encoded* with a different binary number so that the microprocessor can distinguish which keys are being operated in the system.

Displays are used as readouts for a logic system. Seven-segment displays like the one in Fig. 9-10 are very popular, but they have only a limited number of alphabet

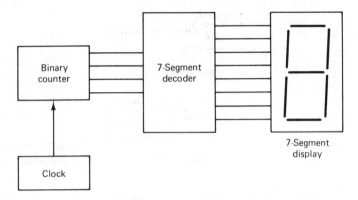

Figure 9-10   Use of a seven-segment decoder to change a binary number to a decimal number.

letters that can be displayed. Alphanumeric symbols are usually delivered to a dot-matrix combination like the one shown in Fig. 9-11. These displays may be either LED (light-emitting diodes) or they may be made with liquid crystals that change the amount of reflected light according to an input voltage.

Another popular display is the gaseous type, which utilizes gas ionization to produce light. As a general rule, you will find that gaseous displays require much higher voltage than either liquid crystal or LED displays.

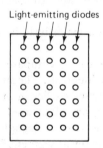

Light-emitting diodes

Figure 9-11   A dot matrix that uses light-emitting diodes for displaying alpha-numeric characters.

Interfaces are used when it is necessary to convert from one logic level to another. An example is in gaseous display. The logic output of a microprocessor might be at a 5-volt level, which would be insufficient to light most gas displays. Therefore, some kind of an interface is utilized so that the 5 volts is actually used to control the higher DC voltage.

A popular interface is the *optical coupler* of the type shown in Fig. 9-12. The input signal produces a light from the LED. That light turns on an output device such as a phototransistor or photo SCR (light-activated SCR, or laser).

The two circuit voltages are isolated from each other because only a light source is used for triggering the output device.

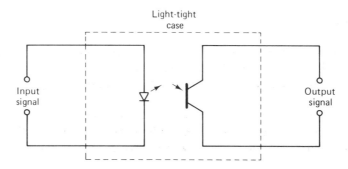

Figure 9-12 One type of optical coupler.

KNOWLEDGE AS A VALUABLE TROUBLESHOOTING AID

As in analog circuits, your knowledge of circuit theory and circuit behavior is one of your most important troubleshooting aids. Certainly, you cannot expect to trouble-shoot an ENABLE circuit if you don't know how an AND gate works. This is carried through to all of the digital circuitry that requires troubleshooting and servicing.

You wouldn't be reading a book on troubleshooting if you didn't know how electronic circuits work. This is especially true of digital logic circuits and microprocessor circuits. However, the brief review of some of the basic components and circuits just given will be useful because knowledge is, after all, a troubleshooting tool.

TYPES OF TEST EQUIPMENT USEFUL IN TROUBLESHOOTING LOGIC AND MICROPROCESSOR CIRCUITS

The first step in troubleshooting any electronic system is to measure the power supply voltage. Next, or at the same time, look for obvious faults.

In logic and microprocessor systems, an accurate voltmeter should be used because the power supply in logic circuits and microprocessor circuits is very often stiffly regulated. Because of the close tolerance of the power supply output, a digital voltmeter is highly desirable for this measurement.

As a rule, voltmeters should *not* be used for measuring logic levels at various points in a system. One reason is that it would be necessary to make a decision between logic 1 and logic 0 in many cases. For example, in a 5V logic system, is 2.8V satisfactory for a logic 1?

To eliminate those decisions, logic probes are much more useful. Another advantage of a logic probe is that you don't have to look up from the work each time you make a measurement. Usually the probe has some type of light indicator so that you can see the level of the logic signal being probed without moving your head.

That second reason may look like begging for an excuse to use a logic probe. However, after two or three hundred measurements with a voltmeter you would very quickly get the idea that a probe is much more convenient.

Logic Probes

A logic probe, like the one in Fig. 9-13, is a very special kind of voltmeter. It is used in digital circuits to indicate the presence or absence of a voltage where that voltage represents a binary number. Very often the presence of a voltage is referred to as logic 1 and the absence is called logic 0.

As mentioned before, it is not a good idea to use an ordinary voltmeter in these measurements because it requires you to make judgments. With the logic probe, the light is either on or off. In other words it is a go/no-go device.

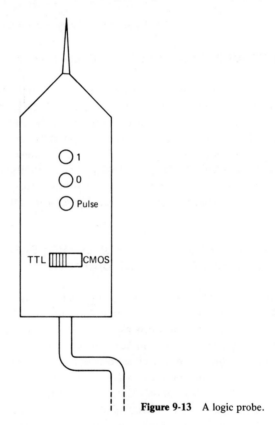

Figure 9-13 A logic probe.

Commercial logic probes have a useful feature called *pulse*. It can be used to determine if there is a presence of pulses or square waves as opposed to a steady DC or zero volts. In some simple logic probes the presence of a pulse will usually cause both lights to light when there is no separate pulse indication. However, if the pulse is of very short duration (low duty cycle) you may get an indication of logic 0 even though the pulse is present.

Another advantage of using the logic probe is that it is possible to catch a very short-duration glitch. A glitch is an undesired spike voltage in a logic circuit. The inertia of a VOM and the time constant of a DMM are usually so long that they do not register the short-duration glitches. However, the glitches are very important because they can cause trouble in a digital or microprocessor system. It may be the very trouble that you are looking for. You would miss that glitch if you didn't use a logic probe.

Some oscilloscopes—especially those having bandwidths of 20 megahertz or less—will miss the very short-duration glitches. So, in this rare case, the logic probe is actually better than an oscilloscope for some types of measurements.

When using a logic probe, always be sure to start by probing the DC supply voltage. This should give you a logic 1, but *it is possible to have glitches on the power supply.* They will be indicated by the pulse indicator on the probe. Remember, glitches can be the cause of the trouble you are looking for.

Logic Pulsers

A logic pulser does the same job in digital systems as a signal generator does in analog systems. It injects a desired signal for the purpose of testing a component or circuit or system.

A combination of a logic pulser and logic probe is sometimes sold at a reduced price.

Figure 9-14 shows a troubleshooting problem where a combination pulser and probe can save time by preventing a wrong interpretation of a measurement.

In Fig. 9-14(a) the logic probe shows a logic zero output or the AND gate. How-

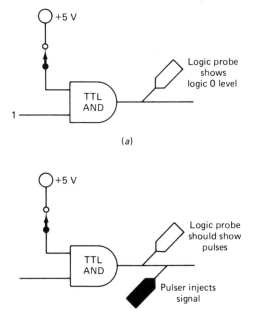

Figure 9-14   How a logic pulser can be used to determine if a gate is not working properly.

ever, you can't tell if this is due to a defective AND gate or if the circuit has been
grounded at this point.

NOTE

> The circuit should have a logic 1 output. You know this from the AND truth table. If the
> circuit is grounded at the output, the AND gate will likely be destroyed. However, if you
> didn't know about the ground you might replace the AND gate and immediately destroy
> the new one.

Figure 9-14(b) shows how to test for this type of problem. The pulser injects the
signal and you use the logic probe to look for that signal. If the problem is with the
gate, the probe will show that a pulse signal is present. If the point is grounded, the
probe will indicate that no signal is present.

Frequency Counters

A frequency counter is a very valuable asset in troubleshooting logic circuits. With this
instrument you can locate clock signals and verify their frequency. Of course, you
could also do the same thing with a good oscilloscope. However, that usually requires
you to make a simple calculation. After tracking the clock signal through several divide circuits, those calculations become tiresome.

It should be mentioned here that some oscilloscopes have frequency counters
built in. They are very useful for this type of measurement because you can count the
frequency and you can also look at the waveform. However, this type of equipment is
expensive, and you have to decide whether the additional expense is justified.

There is also the additional complexity that must be taken into consideration if
the test equipment is down for any reason. In other words, the cost of repairing the
equipment in terms of money and labor will be greater than for individual test
instruments.

A special type of oscilloscope, called the *logic analyzer,* is also useful for troubleshooting. This is a multitrace oscilloscope that enables you to look at as many as six or
eight (or more) logic signals simultaneously. That, in turn, allows you to check for
coincidence and timing of pulses. As with other specialty oscilloscopes, these analyzers
are more expensive than ordinary oscilloscopes. Their added expense must be weighed
against their potential use and the type of troubleshooting you are doing.

MICROPROCESSORS

A good way to think of a microprocessor is to consider it to be a memory operator. This
concept can also be applied to computers.

All of the useful information in a microprocessor (and computer) system is
stored in a memory. The job of the microprocessor is to store, retrieve, and process
information.

Since the microprocessor and computer do the same type of work, the discussion on troubleshooting will be directed toward the microprocessor system. One vast difference between the two systems is in the amount of memory involved. Computers use microprocessors to control a very large amount of memory.

It is very important to know that much of the troubleshooting required for these systems centers around mechanical problems. Plugs and sockets are especially troublesome. The pins become loose and, sometimes, broken.

A loose pin can cause impulse noise and intermittent circuit operation. These are difficult troubleshooting problems. The methods used to locate the sources of these problems are not much different from the methods for linear systems.

The unit connected through the plug will often show symptoms. For example, if the defective plug connects a memory section, that memory will likely be unable to store and return information, at least on the line with the defective pin.

Since only one of the many pins is likely to be a problem, scope each output, one at a time, while putting a varying pressure on the plug. Look for a disturbance on the scope screen.

AN EXAMPLE OF A MICROPROCESSOR SYSTEM

An example of a microprocessor system is shown in Fig. 9-15. This system requires a very stiffly regulated +5V supply. The clock signal is needed for timing all of the operations in the system. Without this clock signal, *nothing* will happen in the system.

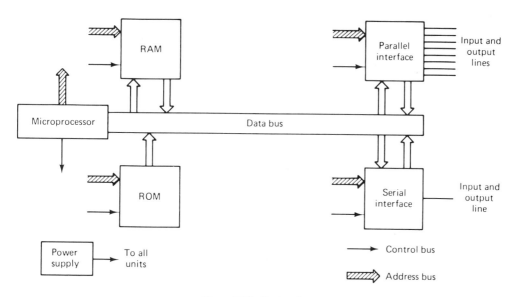

Figure 9-15 Basic microprocessor system.

ROM (read only memory) stores the programs. The programs in ROM are usually put in by the manufacturer. However, there is equipment to program ROMs in the field.

RAM (random access memory) is used by the microprocessor for storing and retrieving numbers. The numbers represent various operations of the system.

The *interface* protects the system from overloads and undesirable external signals.

A *bus* is simply a combination of wires. Often, the bus has 8 wires for data and 16 wires for addresses. The *addresses* are used for locating information in the memories.

The *control bus* carries control signals for operating the various sections. For example, it can set the interface so that information can be delivered to the microprocessor from the outside world. The control signals can also be used to set the interface so that the microprocessor can deliver information to the outside world.

You have already read how a microprocessor can be used in a frequency synthesizer (Chap. 7).

Figure 9-16 shows how a microprocessor is used in a simple four-function calculator. In this case, the microprocessor system contains all of the sections shown in Fig. 9-14. However, they are all in a single integrated circuit. This application is for a dedicated microprocessor.

A *dedicated microprocessor* is one designed and programmed to do a specific job. In this case, that job would be performing four functions (add, subtract, multiply, and divide). This microprocessor cannot be programmed to do some other task. It is the main difference between the microprocessor in Fig. 9-16 and the one in Fig. 9-15.

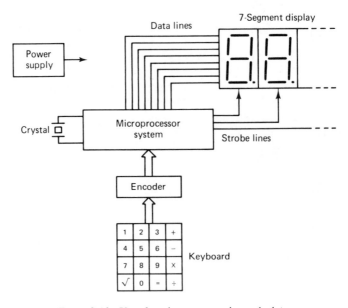

Figure 9-16   Use of a microprocessor in a calculator.

The power supply in this case is a battery—a good starting point for trouble-shooting. The crystal in this particular application is external to the microprocessor system, but in some modern systems the crystal is located internally. The advantage of locating it externally is that crystals sometimes crack and need to be replaced.

However, if the calculator costs less than $100 it is not a serviceable item.

NOTE

In modern electronics any system that costs less than $100 is considered to be nonrepairable. The reason is that the labor for fixing it can be greater than the cost of the system.

Starting with the keyboard we will discuss the simple addition of 2 + 3. The microprocessor continually polls the keyboard. In other words, it continually looks for an input signal from the keyboard.

Your first step in the addition problem is to enter the number 2. When you push the 2 key on the keyboard the microprocessor takes the binary equivalent of that number 2 (which comes out of the encoder) and puts it into a temporary memory inside the microprocessor. Some manufacturers call the temporary memories *registers;* others call them *accumulators*. In either case, the number is entered and stored.

The next keystroke is for the plus sign. When that plus key is pushed, the encoder sends a coded number to the microprocessor. That number tells it to go to ROM and fetch a program for adding two numbers. Think of the program as being a plan for performing a certain job. That program is entered into a special section in the microprocessor where it is stored for future use.

The third keystroke is for the number 3. That number is delivered through the encoder. A binary equivalent of the number 3 is sent to the microprocessor where it is also temporarily stored. In this case, the number is stored in the second accumulator, or another register.

When you push the equal sign key, a coded message from the encoder tells the microprocessor to start the process of addition using the program that was fetched from ROM. The plan for adding the numbers is sometimes referred to as an algorithm. It is simply the procedure that must be followed to get the answer.

Microprocessors have an internal section called the ALU (arithmetic logic unit). The ALU is called on to actually add the two numbers and store the answer in RAM. From there the microprocessor sends the output data through data lines to the seven-segment display.

It may be necessary to use buffers (not shown) on the data lines in order to drive the display. This would be especially true if the display is gaseous or light-emitting diodes. Liquid crystal displays do not require a high voltage for their operation, so the microprocessor may be able to deliver signals to the display directly.

The data lines connect to all of the individual sections, each being part of a seven-segment display. The microprocessor sends a code for the first digit and the strobe line enables the first digit. So the first digit is displayed.

For example, suppose the first digit in the answer is a number 4. Then the microprocessor would deliver power for the proper segments to get the digit 4, and the strobe output would enable the first digit so that it could be displayed. The power for the second number is then sent through the data lines. (Remember, the data lines connect to all of the digits in the display.) Now the second strobe line enables the second digit and it is displayed. This process continues through as many digits as there are in the display.

The numbers are lighted in sequence so rapidly that your eye believes they are on at all times. By lighting the digits one at a time the power supply drain is greatly reduced. That's the advantage of strobing the display.

Troubleshooting a Microprocessor

Disregard the fact that a calculator is probably a nonrepairable device because a calculator of this type can be replaced for a few dollars. Suppose you need to repair it, and you have completed the preliminary checks.

The first step is to isolate the problem.

Assume the problem is that the second digit in the display does not light. That could be because the strobe voltage is not being delivered or because the second digit is defective. If the digits can be replaced singly, the next step is to disconnect the strobe line and see if you can light the problem digit with a separate power supply. If you can light it, the trouble is in the microprocessor and it will have to be replaced.

If you can't get the display to light with a separate supply, it must be replaced. This is usually a very difficult, if not impossible, job. Only a few types of strobbed displays have replaceable elements.

Troubleshooting is a matter of thinking logically, asking yourself, What could be the cause of this trouble? Then your knowledge of the system enables you to determine what procedure is necessary in order to test for that trouble.

Microprocessors are electronic large-scale integrated circuits that are used to control memory. In a microprocessor system or computer, the microprocessor is able to utilize all of the information stored in very large memories. The manufacturer provides special codes for getting the information into and out of memory, operating upon the coded information, and delivering the result to some output destination.

Microprocessors are *dynamic,* so there is always a clock signal associated with these devices. Because of the wide variety of inputs and outputs to microprocessor systems, it is often a good idea to resort to diagnostics for troubleshooting these systems. This assumes, of course, that the preliminary tests do not show a faulty circuit or component. It also assumes that you have tested the mechanical parts when they are a possible problem.

SUMMARY

Many of the troubleshooting procedures for a linear system are the same as for a digital system. The test equipment is different, the logic probe is preferred over a voltmeter, and logic pulsers are used instead of signal generators. Nevertheless, they are used in a similar way for signal tracing and signal injection.

As with linear systems, your best troubleshooting aid is your knowledge of how digital and microprocessor systems work.

In computer systems the best way to locate a defective section is by diagnostics. However, remember that plugs and mechanically operated parts are prime suspects for troubles in computer systems. Check those first.

If you have not already done so, take time to memorize the symbols for the gates and the truth tables.

PROGRAMMED SECTION

The instructions for this programmed section are given in Chap. 1.

---

1. Which of the following best describes the general purpose of a microprocessor?

   A. It uses information stored in memory to accomplish various tasks. Go to block 7.

   B. It is used for operating machinery. Go to block 14.

---

2. You have selected the wrong answer for the question in block 9. Read the question again. Then go to the block with the correct answer.

---

3. You have selected the wrong answer for the question in block 17. Read the question again. Then go to the block with the correct answer.

---

4. The correct answer to the question in block 18 is B. A decoder converts binary numbers into some more convenient output. An encoder converts some convenient input to a binary code.

   Here is your next question:

   An oscilloscope with a relatively narrow bandwidth, such as 20 mHz, is likely to miss a

   A. glitch. Go to block 25.

   B. logic level. Go to block 20.

---

5. You have selected the wrong answer for the question in block 21. Read the question again. Then go to the block with the correct answer.

---

6. You have selected the wrong answer for the question in block 13. Read the question again. Then go to the block with the correct answer.

---

7. The correct answer to the question in block 1 is A. The microprocessor cannot operate without memory. In some cases, the memory is in the same integrated circuit as the microprocessor.

   Here is your next question:

   What are the only logic levels used in logic and microprocessor circuits?

   A. forward and reverse. Go to block 19.

   B. 1 and 0. Go to block 12.

---

8. You have selected the wrong answer for the question in block 12. Read the question again. Then go to the block with the correct answer.

---

9. The correct answer to the question in block 25 is B. There are three important buses in a microprocessor:

Data bus: carries information

Address bus: carries information on locations

Control bus: carries instructions for operating various parts of a microprocessor system

Here is your next question:

A microprocessor that is designed to do a specific job and cannot be programmed is called

A. a dedicated microprocessor. Go to block 26.

B. a single-step microprocessor. Go to block 2.

---

10. The correct answer to the question in block 17 is A. The overbar means to invert. For example, $\bar{A}$ means to invert A. Another way of saying $\bar{A}$ is INVERTED A.

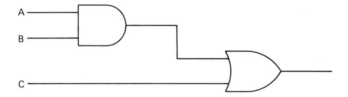

Here is your next question:

The output of the circuit in the illustration for this block should be

A. A OR B OR C. Go to block 22.

B. A AND B OR C. Go to block 13.

---

11. You have selected the wrong answer for the question in block 18. Read the question again. Then go to the block with the correct answer.

---

12. The correct answer to the question in block 7 is B. Having only two logic levels makes troubleshooting easier in many ways.

Here is your next question:

Two of the most important logic families will work with power +5V supplies. Which will *not* work with a +5V supply?

A. CMOS. Go to block 8.

B. ECL. Go to block 23.

C. TTL. Go to block 17.

---

13. The correct answer to the question in block 10 is B. Consult the illustration in this block. Note that the output of the AND gate is A AND B. That is one of the inputs to the OR gate. The other input is C. The two inputs to the OR gate are combined to make A AND B OR C.

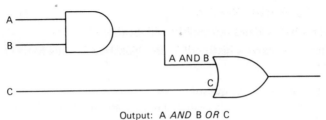

Output: A *AND* B *OR* C

In symbols: (A × B) + C

Here is your next question:

Identify the gate that is sometimes called a logic comparator.

A. EXCLUSIVE OR. Go to block 6.

B. EXCLUSIVE NOR. Go to block 24.

---

14. You have selected the wrong answer for the question in block 1. Read the question again. Then go to the block with the correct answer.

---

15. You have selected the wrong answer for the question in block 25. Read the question again. Then go to the block with the correct answer.

---

16. You have selected the wrong answer for the question in block 24. Read the question again. Then go to the block with the correct answer.

---

17. The correct answer to the question in block 12 is C. The ECL family requires a power supply voltage of −5V. ECL is the fastest logic family. In other words, it has the shortest propagation delay.

Here is your next question:

Which of the following means NOT M?

A. $\bar{M}$. Go to block 10.

B. NM. Go to block 3.

---

18. The correct answer to the question in block 21 is B. The output can be either Q = 1 and $\bar{Q}$ = 0 or Q = 0 and $\bar{Q}$ = 1. When Q = 1 the flip-flop is said to be high, or in a logic 1 state.

Here is your next question:

To convert a keystroke into a binary code, you need

A. a decoder. Go to block 11.

B. an encoder. Go to block 4.

---

19. You have selected the wrong answer for the question in block 7. Read the question again. Then go to the block with the correct answer.

---

20. You have selected the wrong answer for the question in block 4. Read the question again. Then go to the block with the correct answer.

21. The correct answer to the question in block 24 is A. A toggled flip-flop is a *divide by two device.*

      Here is your next question:

Which logic circuit device is sometimes called a bistable circuit?

A. NOR GATE. Go to block 5.

B. flip-flop. Go to block 18.

---

22. You have selected the wrong answer for the question in block 10. Read the question again. Then go to the block with the correct answer.

---

23. You have selected the wrong answer for the question in block 12. Read the question again. Then go to the block with the correct answer.

---

24. The correct answer to the question in block 13 is B. The truth table for the EXCLUSIVE NOR is shown in Table 9-2. Note that the inputs must be the same in order to get a logic 1 level at the output.

      Here is your next question:

A clock signal is delivered to a toggled flip-flop. The output frequency is

A. one-half the clock frequency. Go to block 21.

B. twice the clock frequency. Go to block 16.

---

25. The correct answer to the question in block 4 is A. Oscilloscopes with a bandwidth of 50 mHz are not uncommon in logic system troubleshooting.

      Here is your next question:

A combination of wires that conveys a parallel logic signal (called a *word*) is

A. a logic train. Go to block 15.

B. a bus. Go to block 9.

---

26. The correct answer to the question in block 9 is A. A bus is identified by the number of conductors it has. For example, an 8-bit bus can carry eight binary digits (bits) and it has eight conductors.

      Here is your next question:

Name four things a logic probe can tell you. Go to block 27.

---

27. A commercially manufactured logic probe can tell you if a point in the circuit is at logic 1 or logic 0. Also, it can tell you if a glitch or pulse is present.

You have now completed the programmed section.

TEST YOUR KNOWLEDGE

1. Name three of the most popular integrated circuit families. ____, ____, ____
2. Which gate is used to make an ENABLE? ____
3. Identify the gate that has this truth table. ____

0	0	1
0	1	0
1	0	0
1	1	0

4. What type of gate is used with a −5V power supply? ____
5. What does $\bar{A}$ mean (in words)? ____
6. What are the three possible outputs of a three-state buffer? ____, ____, ____
7. What is the purpose of a register? ____
8. What is the name of the logic device used to deliver a single input to a number of output terminals? ____
9. What are the two outputs of a flip-flop? ____, ____
10. A microprocessor is used to ____.

ANSWERS
TO TEST YOUR KNOWLEDGE

1. TTL, CMOS, ECL
2. AND
3. NOR
4. ECL
5. NOT A
6. logic 1, logic 0, open circuit
7. for short-term memory
8. demultiplexer
9. Q, $\bar{Q}$
10. utilize memory

10

Repairing and Replacing

CHAPTER OVERVIEW

Despite the possibility that it may be viewed as "negative teaching," this chapter has much to say about what you should *not* do when replacing and soldering parts.

Some of this material is taken from bulletins issued by the National Aeronautics and Space Administration (NASA). Other sources are publications supplied to technicians by manufacturers of consumer electronic equipment.

When you troubleshoot a system, you are employing your knowledge and your test equipment to find the cause of the trouble. When you replace a component you are, in effect, putting your signature on the work done. If you do unprofessional work you are saying that you don't care enough about your craft to learn how to do it right.

Doing professional work is not just a matter of pride in your work. It can save you time (and sometimes money) by not having to do the work over again. As a general rule, if you follow the best practices your work will not require rework.

Objectives

Here is a partial list of the topics covered in this chapter:

- The most common fault in professional soldering
- How a cold solder joint can be recognized
- Special techniques required for surface mount technology
- Why strong mechanical connections are not desirable for soldering
- Characteristics of a good soldered connection

SOLDERING TECHNIQUES

How would you classify soldering? Is it an art? Is it a technology? Both answers are correct. The *American Heritage Dictionary* gives several definitions of art. One is "a craft or trade and its methods." Another is "any practical skill." The same dictionary defines technology as "the application of science especially in industry or commerce"—also, the methods and material thus used.

Soldering fits all of those descriptions and more.

The Solder

Soldering is a popular method of connecting circuits. One reason for its extensive use is the low cost of the solder connection. Another reason is its high reliability (when it is done properly).

Most of the solder used today is of the 60/40 type—60 percent tin and 40 percent lead. A somewhat better solder (became it melts at a lower temperature) is the 63/37 variety, which has 37 percent lead and 63 percent tin.

Most combinations of lead and tin become plastic before they actually melt. So they have three states: solid, plastic (mushy), and liquid. Sixty/forty solder is this way, but there is a very small difference between the liquid and solid states.

The 63/37 solder is *eutectic* at a temperature of 370°. That means that it goes directly from the solid to the liquid state without having a plastic intermediate step. That is one of the reasons that 63/37 is often used for replacing parts on newer systems where the components are very small and sensitive to heat. You do not want to have to hold the solder on long enough to go through a plastic state between solid and liquid. Figure 10-1 shows the difference between the 60/40 and 63/37 solders when heated.

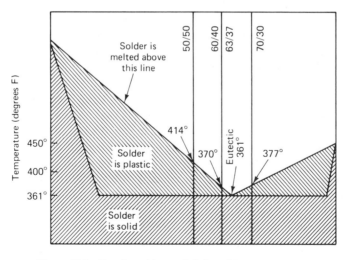

Figure 10-1   Popular solders and their melting temperatures.

Instead of soldering with a combination of lead and tin there are conductive *plastic* solders. They come in a soft state and are hardened by heat or by adding a chemical called a catalyst or hardener. In either case, the plastic solder is applied first and then hardened. In this way the procedure is opposite to soldering with lead and tin, where you start with solid material.

Plastic solders form a very solid strong bond, and they are easy to apply. The one that uses a catalyst for hardening is especially useful in surface mount configurations, where heat must be used sparingly to protect the component.

Some solders (such as silver solder) employ indium. Indium is one of the basic elements. In one type a fuseable alloy of indium, tin, lead, cadmium, and gallium are combined to make a solder. These solders melt at very low temperatures, but they do not flow easily like the more conventional solders. Some of these alloys form solders that melt at so low a temperature that they can be melted with the flame from a match.

WARNING

Cadmium is a *highly poisonous* material! Technicians have died from inhaling solder that contains cadmium. You should *never* inhale any smoke from soldering. In fact, a small fan should be part of your bench equipment, to pull the smoke away from you while you are soldering. You have to inhale a quantity of this material in order for it to kill you, and that is apparently what the technicians did. That is carelessness and it should never have happened.

You should not be overly suspenseful about this warning. If you know what you are doing, you are safe when making solder connections.

Surface mount components do not employ leads like the resistors and capacitors have. See Fig. 10-2. Instead, they have a metal interface that is used for soldering.

Figure 10-3 compares a surface mount component with a conventional component. There are some things you should know about working with surface mount technology. First, you must *never handle the components.* Second, if you *remove* a component from a circuit for testing purposes, you should *never replace that component in the circuit.* Use a new one instead. Otherwise, you are going to be liable for unprofitable callbacks.

One of the best ways to handle surface mount components is to use a vacuum parts holder. It holds the parts by suction so that you do not have to touch them

In some cases these components can be soldered. Also, there is a special solder cream that you can use for soldering onto these circuit boards.

Never use 60/40 solder for soldering surface mount components. The amount of temperature required to melt that solder is surely to destroy the component. As a technician, you should know that solder with an acid flux is *never* used in electronics.

The most convenient solder is the type that has the proper flux (usually resin) in the center or core of the solder. With that type, the flux is applied every time you solder. The flux is needed for cleaning the surfaces in order to make a better solder connection.

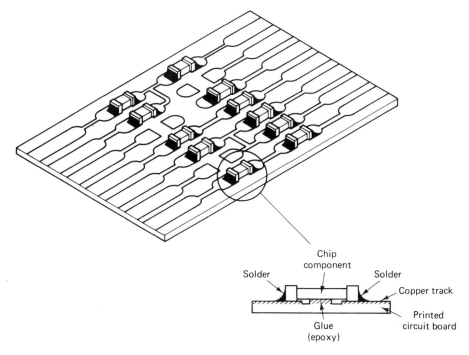

Figure 10-2 A surface mount board and a close-up showing how it is soldered.

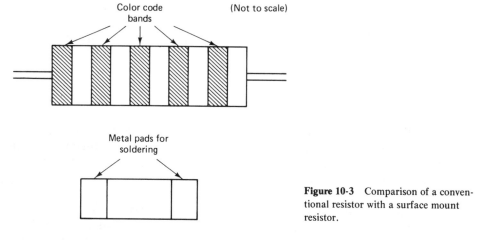

Figure 10-3 Comparison of a conventional resistor with a surface mount resistor.

Some Important Don'ts

Before discussing soldering techniques, it is a good idea to review some of the things that you must *not* do when making a soldering connection.

Some technicians get impatient while they are waiting for the solder to cool and harden. They develop a bad habit of blowing on the solder. *Never* do this. Here are some of the reasons why:

- It causes the solder to cool from the outside in.
- The moisture from your breath is trapped in the solder and that can cause problems.
- At very high frequencies it has been shown that blowing on solder causes high impedance joints.

You do not want to have to stop and think what frequency the circuit is going to be used on before you decide whether to blow on it or not. The best way is to avoid that undesirable practice.

There is another bad practice that technicians seem to get into. For some reason they begin to wiggle the solder while it is cooling. This is a very bad practice because it can produce spaces between the surfaces being soldered. In all cases, it is undesirable to have empty space around the conductor or between the conductor and the solder. It is not clear why some technicians get into this habit, but it is a good idea to avoid it.

Another highly undesirable practice that seems to be handed down by word of mouth is to make a very strong mechanical connection before soldering begins. Consult Fig. 10-4, which shows how to make the connection and how *not* to make the connection.

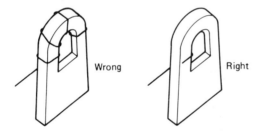

Wrong   Right

Figure 10-4 Comparison of wrong and right ways to prepare for soldering.

The most important reason for not making the mechanical connection is that *it does not serve any purpose.* At one time it was thought that a strong mechanical connection was required for making a good reliable electrical contact. Today it is known that this isn't true. Furthermore, it creates problems.

Mechanical connections make it almost impossible to unsolder the lead or the component once the soldering job has been completed. You may be the one who has to unsolder it in the future when you are troubleshooting the same system. You don't want to make a hard job for yourself; and, you certainly don't want to make a hard job and an unprofessional job for another technician.

Soldering

Figure 10-5 shows terminals ready for soldering. These illustrations should give you an idea of why the mechanical connection is not required.

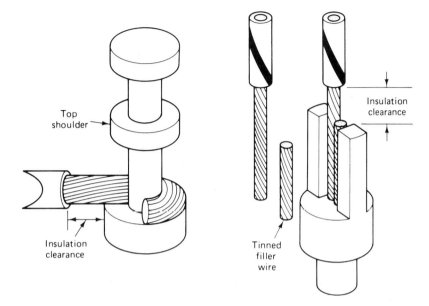

Figure 10-5 Illustrations showing that mechanical connections are not required.

The proper procedure for soldering is shown in Fig. 10-6. Studying these illustrations will not make you good at soldering, but it will at least give you an idea of how to start. You could read ten books on soldering and you still couldn't solder until you pick up an iron and some solder and start making connections.

Be very critical of the soldering you are doing. Practice a sufficient amount of time until you can make a good soldering joint. The most common error in soldering is to use *too much solder*. The second most common is to use *too much heat*.

Too much solder can lead to later troubleshooting problems. Soldering bridges between printed circuit connectors, like the one shown in Fig. 10-7, often results from using too much solder. That, in turn, results in dropping blobs of solder when you are making the solder connection.

Too much heat can destroy the component that is being soldered. In some cases, if sufficient heat is used, it can destroy the printed circuit board itself. Replacing a printed circuit is nearly impossible on large complex electronic systems. So you should

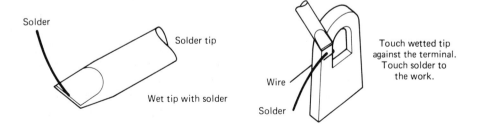

Figure 10-6 The basic soldering procedure.

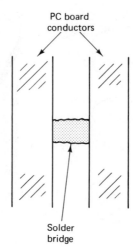

Figure 10-7   A solder bridge.

avoid that necessity by being careful about how much heat you apply and where you apply it.

Another common problem in soldering is to apply too little heat. That causes a dull gray solder connection. It should be shiny.

The dull gray solder connection from too little heat is called a *cold solder joint*. It is unreliable and often has a high resistance. In fact, the resistance may be so high that it acts as if no connection has been made.

Reheating a cold solder joint usually converts it into one that is acceptable.

Under no condition should you ever use solid wire for making electronic circuits. Solid wire is used for making breadboard circuits, but, when you go from the breadboard to the final product, you should never use solid wire.

Solid wire has a tendency to *work-harden*. If you bend it back and forth several times, it becomes very hard and brittle and will break off easily. Stranded wire, on the other hand, is flexible and much easier to work with. A joint made with stranded wire will be more reliable because it does not work-harden like the solid wire.

Inspecting the Solder Connection

As mentioned before, be very critical when you inspect your soldering connection. Practice until you can do it right. If the solder connection is made properly, the solder will be shiny rather than a dull gray cold solder joint.

Another characteristic of a good solder connection is that you should be able to see the strands and the outline of the wire when the solder has cooled. If you put so much solder on that you can't tell where the wire ends and where the terminal begins, you have used too much solder. Remove it and start over.

It is not uncommon for some of the flux to remain on the surface of the solder after it is cool. Always take a cotton tip soaked in alcohol and remove that flux. It won't hurt anything except that it might gather dust particles and make the terminal unsightly. Also, it is not professional to leave the soldered terminal in that condition. It only takes a second to wipe the flux away.

The Soldering Iron

One of the problems that occurs during soldering is that the tip of the soldering iron becomes dirty after you have soldered a number of connections. Some companies make a soldering iron holder with a built-in sponge. They advise you to dampen the sponge and wipe the soldering tip on it frequently to make sure that the tip is clean. If you don't do that, your solder connection will have bits of dirt in it. That is not the signature you want to leave on your work.

Two kinds of soldering tips are available. One is a *solid metal* tip. The manufacturer says it is OK to file those tips occasionally to remove oxidation so that you get a good, clean, heated iron when you make your soldering connections.

The other kind of tip is *plated*. This type is usually recognized because the plating is shiny, especially when it is new. *You must never file this kind of tip!* If you do, you will remove the plating and be in the market for a new tip for your soldering iron.

Unsoldering Connections

You can't put in a new component until you get the old one out. The trick is to get the component out without destroying the circuit board and the terminals. That means you must be very careful when unsoldering.

One of the best techniques is to use a solder sucker, a rubberized flexible bulb that sucks up the solder through a tube. Another type of solder sucker uses a spring-loaded piston.

By using the solder sucker you can remove all of the melted solder. Then the component or the connection will be easy to remove.

Instead of a solder sucker, many technicians prefer a soldering wick. This is a braided wire that removes the solder by a capillary-type action. The procedure is shown in Fig. 10-8. You simply insert the end of the braid into the melted solder and the solder runs up the braid. Then you cut the end off and the braid is ready for use again.

It may sound expensive because you have to keep snipping off the the end of the braid, but, actually, braid is inexpensive. The soldering wick is sometimes preferred because it gives you slightly more control over the place you are removing the solder. You can even shape the end of the braided wire to remove solder from small areas.

The braid should be about the same width as the conductor from which you are removing the solder. Do not use a solder wick that is too wide or too narrow. Some technicians keep several sizes on hand.

After you have removed the solder and removed the component or wire, the next step is to make sure the terminal is clean before you start to replace the wire or component. That may require additional heating with the iron. Again, be careful that you do not overheat and destroy the board or the components.

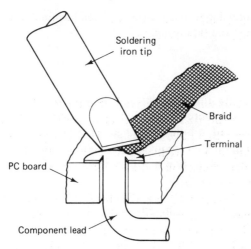

Figure 10-8   Using a wick to remove solder.

Some Additional Soldering Tips

Before you apply the soldering iron to the connection to be soldered, be sure to "tin" the iron. This simply means to melt solder over its surface. The reason for doing this is that the melted solder then forms a good interface. It gets the heat to the place where you want it when you are soldering. Also, make sure the component leads or wire are tinned so that the solder is already fused with the metal. Again, tinning is accomplished by melting solder onto the surface. That way you do not have to hold the iron on a long time in order to get the solder flowing and to complete your connection.

Technicians who do professional work prefer to have a soldering iron with a controlled temperature. They are more expensive, but they permit you to set the temperature to a fixed value. This is especially important when soldering very small components.

The type of soldering gun shown in Fig. 10-9 should *never* be used for soldering modern electronic circuits. It is nearly impossible to control the heat in this type of iron, even with the soldering guns that have the two-step trigger for two levels of heat.

Another reason for not using this type of soldering gun is that the secondary of this gun is insulated from the primary and from other conducting paths. Thus, static

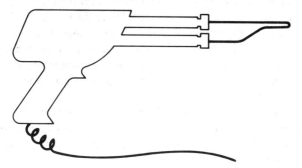

Figure 10-9   Do not use this type of soldering gun.

charges can build up on these tips that can readily destroy a MOSFET or other field effect device. Figure 10-10 shows the schematic of the soldering gun.

No matter what kind of iron you use, always make sure that it does not have static charges on it, especially if you are working on electronic systems.

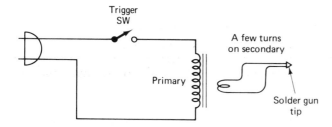

Figure 10-10 This is the circuit for the soldering gun in Fig. 10-9. Note that the secondary is isolated so that electrostatic charges can accumulate and destroy MOSFETs.

When you are soldering heat-sensitive components—that is, components that can be easily destroyed by heat—it is a good idea to use a heat sync of the type shown in Fig. 10-10. The theory is that the heat will flow up the lead and into the heat sync rather than into the component.

Although this next point is not directly related to soldering, it is a problem that you should address as far as transistors and other semiconductors are concerned. It has been shown that when you use cutting pliers on the leads of a transistor or other semiconductor device, it is possible for a shock wave to travel up the conductor and destroy the component internally. Avoid this by using your finger as a shock absorber between the cutters and the component.

Much has been said about the destruction of MOSFETs by handling. The destruction is due to static electricity. Many of the newer MOSFETs and integrated circuit MOS devices are buffered in such a way that static voltages are routed around the components, so the problem of static electricity destruction has been greatly reduced. Nevertheless, you should employ reasonable precautions when working with these devices. Keep the leads together before they are soldered and make sure that the tip of your iron is grounded.

SUMMARY

You can't learn soldering from a chapter in a book, but you have reviewed some of the things to do and some things *not* to do.

You have also reviewed the characteristics of a good solder connection. The appearance of a cold solder joint and how it occurs were also discussed.

Types of soldering irons and tips on their care were discussed, as well as types of solder, including plastic solders.

To serve as guides, pictures of soldering were shown in this chapter. However, you won't become good at soldering unless you practice.

Your work is your personal signature.

PROGRAMMED SECTION

The instructions for this programmed section are given in Chap. 1.

---

1. When soldering in electronic circuits,
   A. always use acid flux to clean the surfaces. Go to block 7.
   B. never use acid flux to clean the surfaces. Go to block 14.

---

2. The correct answer to the question in block 6 is B. You can file the tips if they are solid but not if they are plated.
   Here is your next question:
   If a solder goes directly from the solid state to the liquid state, it is said to be
   A. eutectic. Go to block 15.
   B. electric. Go to block 5.

---

3. The correct answer to the question in block 11 is B. It is a matter of opinion whether you use braid or a solder sucker.
   Here is your next question:
   Is the following statement correct? After soldering a wire to a terminal, you should be able to see the outline of the wire and strands of the wire. Go to block 16.

---

4. You have selected the wrong answer for the question in block 8. Read the question again. Then go to the block with the correct answer.

---

5. You have selected the wrong answer for the question in block 2. Read the question again. Then go to the block with the correct answer.

---

6. The correct answer to the question in block 14 is B. Never move the conductor when the solder is cooling. If you do, you may create pockets between the conductor and solder terminal.
   Here is your next question:
   Which of the following statements is correct?
   A. Never file the tip of a soldering iron. Go to block 12.
   B. It is OK to file the tips of some types of soldering irons. Go to block 2.

---

7. You have selected the wrong answer for the question in block 1. Read the question again. Then go to the block with the correct answer.

---

8. The correct answer to the question in block 15 is B. As a general rule, you should make it a practice to avoid inhaling anything but air.
   Here is your next question:
   While waiting for a soldered connection to cool, you should
   A. not get into the habit of blowing on it. Go to block 11.

B. blow gently on it to hasten the cooling. Go to block 4.

9. You have selected the wrong answer for the question in block 11. Read the question again. Then go to the block with the correct answer.

10. You have selected the wrong answer for the question in block 14. Read the question again. Then go to the block with the correct answer.

11. The correct answer to the question in block 8 is A. The best way is to let the solder connection cool without your help.
    Here is your next question:
    Instead of a solder sucker, some technicians prefer to use
    A. stranded wire. Go to block 9.
    B. braid. Go to block 3.

12. You have selected the wrong answer for the question in block 6. Read the question again. Then go to the block with the correct answer.

13. You have selected the wrong answer for the question in block 15. Read the question again. Then go to the block with the correct answer.

14. The correct answer to the question in block 1 is B. Acid flux is highly corrosive and will destroy electronic parts and connectors.
    Here is your next question:
    Is the following statement correct? While waiting for solder to cool, you should wiggle the conductor. That way you can tell when the solder has hardened.
    A. Correct. Go to block 10.
    B. Incorrect. Go to block 6.

15. The correct answer to the question in block 2 is A. An alloy solder is eutectic at only one temperature.
    Here is your next question.
    The fumes from molten solder are highly poisonous if the solder contains
    A. tin. Go to block 13.
    B. cadmium. Go to block 8.

16. The correct answer to the question in block 3 is yes. Remember that too much solder is a typical problem with solder connections.
    Here is your next question:
    How do you remove flux that has not burned off during the soldering process? Go to block 17.

17. Remove the flux with alcohol and a cotton swab. Do not use any other type of cleaning agent because some of them leave deposits that attract dirt.

You have now completed the programmed section.

TEST YOUR KNOWLEDGE

1. Name two ways that plastic solders are hardened. ____, ____
2. Which types of components should never be reconnected into the circuit after they have been removed? ____
3. If a solder goes directly from the solid to liquid state when heated, it is said to be ____.
4. Which type of flux is *never* used to solder electronic components and circuit? ____
5. Should a strong mechanical connection be made when soldering a conductor to a terminal? ____
6. Two common mistakes in soldering are too much ____ and too much ____.
7. What is the cause of a cold solder joint? ____
8. To remove flux from a solder connection after it has cooled, use a cotton swab and ____.
9. Which type of soldering tip should *never* be filed? ____
10. What is a solder wick made from? ____

ANSWERS
TO TEST YOUR KNOWLEDGE

1. heat, catalyst or hardener (the method depends on the type of plastic solder)
2. surface mount
3. eutectic
4. acid flux
5. no
6. solder and heat
7. not enough heat during soldering
8. alcohol
9. a plated tip
10. braided copper wire

11

Some Low-Cost

Homemade Testing Devices

CHAPTER OVERVIEW

Your ability to troubleshoot depends directly on your ability to use manufactured test equipment. There are, however, some simple test instruments that you can build yourself. Some of those instruments have been shown in previous chapters (but not identified as homemade).

When is it a good idea for students and technicians to construct test equipment? Here are a few examples:

- When the test equipment cannot be purchased because it isn't being manufactured. (One reason it isn't being manufactured is that it is so simple that it cannot be priced profitably.)
- When the test equipment is needed for certain specialized tests.
- When simpler, less expensive test equipment is the only alternative until more funds are available.
- Because it is interesting to make (and try) electronic projects.
- For the learning experience.

The examples chosen for this chapter are intended to illustrate constructed test equipment. You can find many other examples in magazines and books. These projects are simple. It doesn't take much time to put them together.

You may be disappointed to find that construction is not described down to the finest detail. This allows you to use the things you have on hand rather than purchase parts and supplies. Also, much of the fun of building a project is in laying out the work.

Despite the many positive ideas about building test equipment, it is still a good idea to use manufactured equipment whenever possible. This is especially true if you are working in a position where your activities are being observed by people who depend on, and except, your professional expertise, for example, owners of expensive equipment that you are servicing. Those customers will not be impressed (usually) by the fact that you made the test equipment yourself.

Objectives

Here is a partial list of the projects included in this chapter:

- Battery pack (used for bias and for defeating closed loops)
- Signal injector
- Demodulator probe
- Resistor substitution box
- In situ transistor tester
- Adapter for measuring AC current

NONPOLARIZED ELECTROLYTIC CAPACITORS

Because electrolytic capacitors are polarized, they can only be used in DC and pulsating DC circuits. However, if you are making your own projects, or if you have some special application, it is important to have a very large capacitance that can be used in an AC circuit.

Figure 11-1 shows how a nonpolarized capacitor can be made. The diodes (Da and Db) are called *steering diodes*. The broken arrows show the electron current flow in this circuit when A is negative with respect to B. In other words, if B is common this is the negative half cycle. Note that the electrons flow through Da and then into and out of Ca during this half cycle.

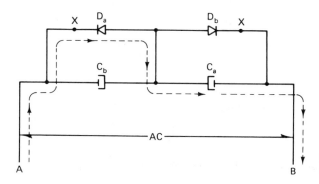

Figure 11-1 A nonpolarized electrolytic capacitor.

When the voltage reverses so that A is positive with respect to B, the current will flow through Db, into and out of Cb, and then to A.

Make sure the diodes can handle the current for the application you have. Remember that the *charging current* for an electrolytic capacitor (like Ca and Cb) is very high and can be in the order of hundreds of amperes. For that reason you may need a surge-limiting resistor in series with the diodes. If so, add a series resistance of less than 100 ohms at the points marked with an X in the schematic. Also, it is a good idea to have a fuse in series with terminal A or B to protect the components in case of heavy surges from the AC source.

Use heavy-duty rectifier diodes with a high current rating.

VOLTAGES TO CHECK CALIBRATION

If you are using a triggered sweep scope, you are sure to have an internal calibrator that permits you to read voltages of waveforms on the display. Likewise, if you are working in electronics, you surely have a multimeter that can read voltages.

However, in both cases it is a good idea to check the calibration of these instruments periodically. Figure 11-2 shows a simple zener diode circuit with three voltages for calibration.

The circuit is very simple to design. You can follow the design procedure given here for making any open-loop regulated supply. For example, if you want a 9V source to replace or substitute for 9V batteries in equipment, you can use the design procedure that will be given.

All of the zener diodes in this circuit should have the same current rating. So, if they have identical voltage ratings, it follows that the power rating (which is usually given) should be the same for each zener.

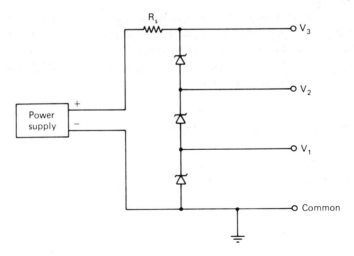

Figure 11-2　A simple voltage calibrator.

The power supply voltage should be at least 50 percent higher than the highest voltage at point V_3. So, if the zener diodes are rated at 5V each, then the voltage at output terminal V_3 will be 15 volts. The power supply voltage should be the value that is calculated as follows:

$$\text{Zener (total) voltage} \times 1.5 = \text{Input power supply voltage}$$

If zener diode total voltage (at point X in Fig. 11-2) is 15V:

$$15V \times 1.5 = 22.5$$

Now you know the power supply voltage and you know the voltage at V_3. So the voltage across R_s is the difference between those two voltages, or $7\frac{1}{2}$ volts. If you disregard the current flow in the external load, then the resistance of R_s can be calculated by Ohm's law. Here is the calculation:

$$\text{Resistance of } R_s = \frac{\text{Voltage across } R_s}{\text{Zener current rating}}$$

As a rule, you can disregard any current flow at the terminal because you are using this to calibrate high-impedance oscilloscopes and meters. If, however, there is current flow in the external circuit, that current must be added to the zener diode current flowing through R_s.

If you have a bench power supply, set it to the voltage input value calculated and try the circuit. Then package it in a convenient box with external terminals.

Don't forget that checking the calibration of your scope and voltmeter can easily be done with a dry cell. The 1.5 terminal voltage of a dry cell remains the same even though its internal resistance is due to old age.

Of course, if you try to draw current from an old dry cell, the voltage drop across the internal resistance will subtract from the 1.5 volts. That would give you an unreliable terminal voltage. Remember that oscilloscopes and voltmeters have very high input impedance. So you can disregard the internal voltage drop of the cell.

As a rule, if the calibration of $1\frac{1}{2}$ volts is correct, all of the calibration of the instrument can be relied upon. The circuit of Fig. 11-2 gives better choices.

USING A VOLT-OHM-MILLIAMMETER FOR A HIGH-VOLTAGE MEASUREMENT

Figure 11-3 shows an example of a series-resistance voltage divider. It can be used to drop a high voltage to a lower value that can be measured with your VOM. The proportional method of calculating the voltage across each resistor can be used as follows:

$$\text{Output voltage} - \text{input voltage} \times \frac{R_5}{R_1 + R_2 + R_3 + R_4 + R_5}$$

Note that the voltage divider shown in Fig. 11-3 divides the voltage to one-tenth

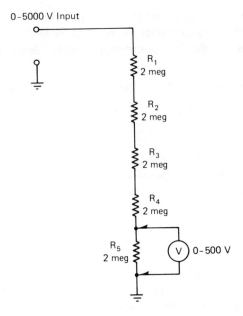

0-5000 V Input

R_1 2 meg

R_2 2 meg

R_3 2 meg

R_4 2 meg

R_5 2 meg

V 0-500 V

Figure 11-3 A voltage divider that permits the measurement of high voltages.

its ordinary value. Therefore, measuring across R_5, the 0 to 500V range is an easy matter. You simply multiply the voltage rating by 10 to get the actual input voltage.

Here it is presumed that 5000V is the absolute maximum voltage. It is safe to use the 500V scale on the meter.

This kind of voltage divider is very easy to design. It is basically the same as the voltage divider used in high-voltage probes for measuring the second anode voltage of CRTs in television receivers and the cathode voltage in the oscilloscope. (In an oscilloscope, it is common practice to hold the second anode voltage at common, or near common, voltage and operate the CRT with a high negative voltage on the cathode.)

A VERY SIMPLE LOGIC PROBE

Figure 11-4 shows a logic probe that is easy to construct from a discarded pen. An LED is used as an indicator and a series resistor limits the current flowing through the LED.

In this particular case, it is assumed that a logic probe will be used in circuits where a logic 1 is +5 volts and a logic 0 is 0 volts.

You can make this probe compact and carry it with you, but remember that it is always better to use a professional logic probe when one is available. The homemade probe is only used for a quick troubleshooting procedure.

Professional logic probes can tell you whether there is a logic 0 present or an open circuit, something that is not possible with the simple probe in Fig. 11-4. When the LED doesn't light, you don't know if the circuit is open or if you are probing a

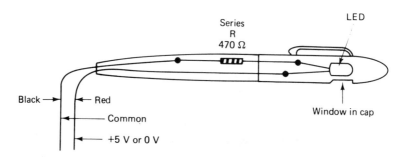

Figure 11-4 A very simple logic probe.

place where the circuit is grounded. Also, the probe in Fig. 11-4 cannot tell you if a glitch or pulse is present. A professional logic probe can do those things.

A SIMPLE NOISE GENERATOR FOR USE IN SIGNAL TRACING

You will remember that in signal tracing a signal generator of some type is used at the input. You follow the signal through the circuitry (using an oscilloscope or voltmeter). In this case, it will be assumed that you are using an oscilloscope.

The circuit of Fig. 11-5 shows a simple noise generator. Note that the silicon diode is reversed biased so that you are producing a small leakage current in the re-

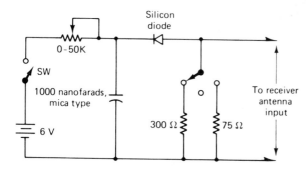

Figure 11-5 An example of a noise generator.

verse direction. (Incidentally, germanium diodes work very well in this application.) The reverse current through the diode produces a noise output.

The output test leads are delivered to the antenna terminals of the high-frequency receiver. Then, using an oscilloscope, you follow the noise signals through the r-f and detector sections. As soon as you come to the point where you can't see the noise signal anymore, you have passed the cause of the system being inoperative.

When you first connect the circuit to the antenna terminals, adjust the variable resistor for maximum noise at the output of the first r-f amplifier. The receiver should be tuned to a frequency where there is no station.

The 300-ohm and 75-ohm terminals are most common for television and radio receivers. Note that there is a switch position without a resistor. The resistors are only needed if there is an isolation capacitor inside the receiver in the antenna line.

Instead of using an ordinary silicon diode, you may be able to obtain a noise diode, which is specifically made for the purpose of generating noise. That, of course, will give you better results.

MEASURING AC CURRENT WITH A VOM

Most voltmeters and volt-ohm-milliammeters are unable to measure an AC current, especially a relatively high current.

The setup in Fig. 11-6 is convenient when you have to make the AC current measurement. It consists of two terminals for connecting the AC lead to the power

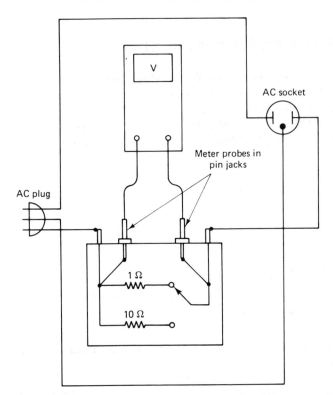

Figure 11-6   A method of measuring AC current.

plug and two pin jacks that you can use for your VOM leads. There is a 1-ohm and 10-ohm resistor in series with the AC current being measured.

Remember that if a 1-ohm resistor is used, then the voltmeter will directly display the current flow. If you use the 10-ohm position, you must divide the voltmeter reading by 10 to get the value of current.

It is very important that you use wire-wound high-power resistors and allow plenty of room for ventilation if you construct this device. The output plug is used as a receptacle for the system in which you are measuring the current.

Everything in the circuit of Fig. 11-6 is constructed except the meter and meter probes. With the finished instrument you simply insert the AC plug into the power and plug the system (for which you are measuring the current) into the socket. Connect the meter, set the switch, and measure the AC current.

A RESISTOR SUBSTITUTION DEVICE

You are probably familiar with the use of decade boxes. They are useful for building new circuits and for troubleshooting.

Decade boxes are made with both resistors and capacitors. The resistor type enables you to select a desired value of resistance or to vary that resistance over a range of values in order to get an ideal ohmic or capacitive value for the circuit. Figure 11-7

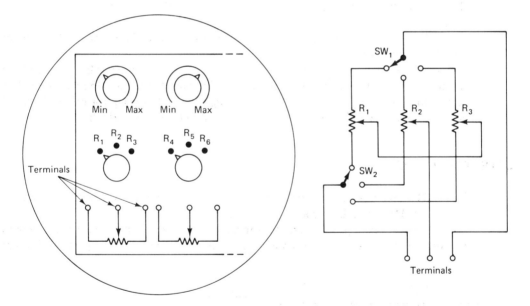

Figure 11-7 A resistance substitution box.

shows an alternative to the resistor decade box. In this case, variable resistors are used. (In a decade box there is a separate resistor for each step.) The schematic shows the connections for one of a number of variable resistors in a box. The cutaway view shows a suggested way of laying these out on top of a box.

The variable resistor in the cutaway view is an illustration printed on the box. The symbol shows the user how the terminals are connected internally. Table 11-1

TABLE 11-1

(Use linear variable resistors)	
Resistor	Range of resistance
R_1	0–500
R_2	0–1K
R_3	0–5K
R_4	0–10K
R_5	0–50K
R_6	0–100K
R_7	0–250K
R_8	0–500K
R_9	0–1 meg
R_{10}	0–2 meg
R_{11}	0–5 meg
R_{12}	0–10 meg

shows the values of the variable resistors in the schematic as well as for two more ranges.

By properly connecting the terminals, you can get just about any combination of resistance value that you want.

One thing should be made clear: A variable resistor should not be operated with the arm close to either of the ends. If that happens you should go to the next step. The reason for this is that it sometimes places high currents through a relatively small value of resistance and the potentiometer is unable to dissipate the heat.

DETECTOR PROBE

Figure 11-8 shows a simple schematic for a detector probe that you can easily make. The diode (D) should be a point contact type.

Point contact types are actually a form of hot carrier diode. They are made with an interface of metal against a semiconductor material. The advantage of this kind of

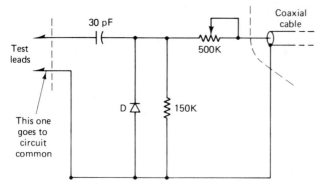

Figure 11-8   A demodulator probe.

diode is that it has a very low forward breakover voltage. Therefore, there will be an output signal at the coaxial cable even though the input signal strength is very low.

One way to build this probe is to use a small box to enclose the parts that are between the broken lines. That way you can use test leads without having the cumbersome probe. You should recognize that the best way to build this would be to construct it in some kind of small case so that the circuit is very close to the end of the test leads. That minimizes distributive capacitance.

If you probe in a circuit that has a modulated signal and the coaxial cable goes to an oscilloscope, you should see the modulation only. If the output goes to a voltmeter, then you will see a deflection that is proportional to the average value of the modulation signal. In either case, you will be able to probe the output even though the r-f frequency is too high for the instrument you are using.

Figure 11-9 shows an example of a circuit that produces a rectangular output.

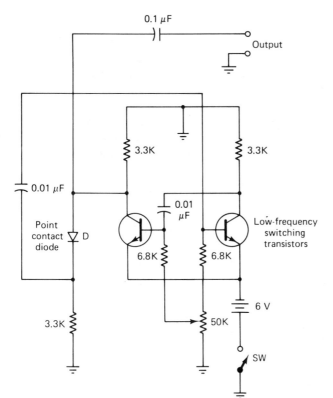

Figure 11-9 An example of a signal injector (any pulse or square wave generator will do).

This circuit can be used as a signal injector in circuits over a wide range of frequencies. The harmonic content of the signal is very high, so it works for both audio and r-f signals.

The signal is produced by a multivibrator. The output signal is taken from the collector of one of the multivibrator transistors. The variable resistor permits you to adjust the frequency in a duty cycle over a range of values.

This is not the only way to make a signal injector. Some technicians prefer to use a simple 555 integrated circuit timer. It can be operated as an astable multivibrator and it will also produce a pulsed output. You might try a 555 circuit as an astable multivibrator instead of the one in Fig. 11-9.

A SIMPLE IN SITU (ON-SITE) TRANSISTOR TESTER

This circuit was suggested by Mike Olszewski Smith of West Palm Beach, Florida. It has been tested for both voltage amplifier and power amplifier transistors.

In situ testers can be used for checking transistors out of a circuit and checking them while wired into a circuit. The theory of this tester is that the transistor, when used in an oscillator circuit, must be good if it supports oscillation. Remember that an oscillator is just an amplifier with regenerative feedback. So, when this device checks a transistor, it is used as the amplifier.

Figure 11-10 shows an example of a simple in situ tester that you can easily build. In this circuit the oscillator is in a Hartley configuration. The primary of the

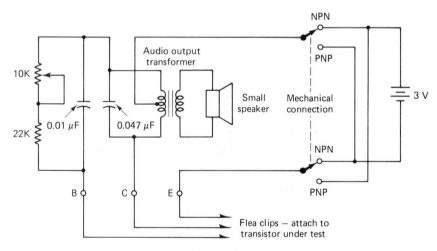

Figure 11-10   An in situ tester.

output transformer is used to shift the phase for feedback, and the 0.047 microfarad capacitor is used to set the audio frequency.

The feedback circuit is through the 10K and 20K resistors. The capacitor across those resistors ensures that high-frequency transients will not produce regenerative feedback and cause the amplifier to oscillate at a high frequency.

Note that both NPN and PNP transistors can be tested. An audio sound from the speaker indicates that the transistor under test is good.

If you are checking a transistor while it is in the circuit, be sure that the power supply for that circuit is disabled. Usually this is accomplished simply by turning off the system.

Index

A

Accuracy, 66
  current meter check, 66
  full scale, 69
  maximum, 69
  of measurement, 68
  voltmeter check, 66
  of voltmeter measurement, 69
Active high, 198
Active low, 198, 200
AFPC, 168
AGC. *See* AVC/AGC
AGC/AVC, 127
Amplification, 26
Amplifier, 37
  audio, 39, 42
  class A, 39, 75, 104
  class AB, 39, 52, 55
  class B, 39, 104–5
  class C, 104
  differential, 37–38
    common-mode, 38
  direct-coupled, 37, 56, 81
  distortion, 107

  emitter follower, 37
  gain, 76
  high-frequency, 104
  linear, 42, 104
  nonlinear, 42
  operational, 38, 40
  perfect, 177
  phase splitter, 39
  power, 38–40, 77
  push-pull, 38–40
  signal, 123
  single-ended, 38
  transistor-sensitive, 138
  troubleshooting, 123
  voltmeter test, 75
Amplifying
  classes, 27
    class A, 28
    class AB, 52, 55
    class B, 28
    class C, 28
  devices, 19–21, 28
Automatic frequency control (AFC),
    127, 165–66
  block diagram, 166

AVC/AGC, 42, 153–57, 159, 170–72
  bias voltage, 55, 157
  clamping, 155
  defeat, 157
  gain of receiver, 177
Average value, 67, 111
Avogadro's law, 187
Avogadro's number, 180

B

Bandwidth, 104
  on power curve, 94
  on voltage curve, 94
Batteries and cells, 70
  dry cell, 66
    carbon-zinc, 66
  pack, 157–59, 170
  terminal voltage, 70
Beta
  AC, 48
  checker, 71
  DC, 48
  test, 48, 50
    checker, 48
    dynamic tests, 48
Bias
  forward, 32
  gate, 53
  long-tail, 30, 41, 47
  methods, 31
    AGC, 31–32
    AVC, 31–32
    battery, 31
    supply, 31
  reverse, 22
  self, 32–33, 51
  simple, 34
  voltage divider, 33–34
Blanking control, 93
Boltzmann's constant, 180, 187
Brownian motion, 178

C

Capacitance
  grid-to-plate, 24
  junction, 23
Capacitors, 43
  charging current, 235
  electrolytic, 6, 12–13, 15–16, 73
  filter, 6
  leakage, 12, 73
  nonpolarized electrolytic, 235
Catastrophic failures, 152
CET test, 64
Clipping, 47
Closed-loop circuits, 153 (*See also*
    AGC/AVC, Automatic frequency
    control, and AVC/AGC)
  frequency controlled, 152
  voltage controlled, 152
Component
  failures, 3
Computer analysis, 140
Configurations
  common, or grounded, 29, 30–31
    base, 29
    cathode, 29
    collector, 29
    drain, 29
    emitter, 29
    follower, 29–30
    gate, 29
    grid, 29
    plate, 29
    source, 29
Control electrodes, 20
Coupled circuits, 35
  direct, 35–36
  impedance, 35–36
  R-C, 35
  transformer, 35–36
Current
  base, 20
  collector, 20

drain, 32
intrinsic, 178–79
source, 32

D

Depletion region, 23, 26
Diagnostic, 138–39
Diodes
    constant current, 38
    hot carrier, 242
    junction, 102
    steering, 235
    varactor, 23
    zener, 236, 347
Diode test, 80
    germanium, 80
    silicon, 80
Dip meters, 164
Distorted output, 3
Distortion, 2, 39, 47, 67
    crossover, 39
    frequency, 103
    in sine wave, 109
    intermodulation, 108–9
    linear, 103
    nonlinear, 108
    phase, 103
Drain, 26
Duty cycle, 206

E

Electronic switch, 94
External sweep, 92

F

Feedback
    negative, 33–34
    regenerative, 47

Fidelity
    maximum, 156
    versus sidebands, 156
Flow chart, 138–39
Form factor, 67
Frequency counter, 99, 208
Frequency measurement, 106
Frequency synthesis, 109
Function generator, 67
    VCO input, 125–26
Fuse, 3
    open, 4
    resistor, 6
    test, 3

G

Gain, 37
    bandwidth, 47
    bipolar transistor, 34
        versus input resistor, 34
    r-f and i-f amplifiers, 153–55
Gate(s), 26, 195–99
Glitch(es), 9, 206
    oscilloscope display, 96
    persistence, 96
Graticule, 100, 126 (*See also*
        Oscilloscope)
Grounded. *See* Configurations

H

Heat sink, 38
Heterodyne, 42
    heterodyning, 52

I

IGFET, 26
Inductive kickback, 74
In situ tester, 141-42, 244

Integrated circuits, 47
Intermittents, 152, 176
  closed-loop circuit, 186
Intrinsic current, 179

J

JFET, 22, 26, 28, 80, 83
Junction
  bias, 22
  capacitance, 23

K

Kelvin (degrees), 180

L

Leakage current, 13
Level shifting, 36
Logic devices, 200–203
Logic families, 195, 216
  ECL, 216
Logic pulser, 207

M

Measurement
  AC current, 63, 240
  efficiency, 7, 11
  qualitative, 7, 8
  quantitative, 7, 8
  square wave, 8
Meter
  analog, 16, 61–63
  digital, 63
  DVM, 63

ESR, 16
input impedance, 63
input resistance, 65
jeweled, 68
loading, 66
microprocessor controlled, 63
mirrored scale, 61
multimeters, 63
ohms-per-volt, 64
parallax, 61, 101
probes, 69
resistance, 66
sensitivity, 65
taut-band, 68
turnover, 67
VOM, 63
Meter movements
  analog, 64
  jeweled, 61
  taut-band, 61
Microprocessors, 208
  ALU, 211
  busses, 210, 215
  clock signal, 209
  dedicated, 210
  interface, 210
  memory, 210
  system, 209
  troubleshooting, 212
Milliammeter, 4
  connections, 13
  use of, 15
Mirrored scale, 61
Mixing, 52
Models, 25–26
Modulate, 42
MOSFET
  amplifier, 51
  bias, 22
  depletion, 27, 51
  dual gate, 24
  enhancement, 22, 51, 53
  gate, 26
  voltage, 22
Multivibrator, 243

N

Noise, 179
  amplifier, 180
  antenna, 179
  in bipolar transistors, 183
  example, 181
  FET, 188
  in field effect transistors, 184
  flicker, 182
  generator, 239
  i-f, 176, 182, 188
  impulse, 185
  Johnson's, 181, 189
  output, 186
  partition, 182-83
  power, 181
  problem, 185
  thermal, 181
  thermal agitation, 181
  white, 181
Noise resistor, 177
Nonsinusoidal voltages, 60

O

Offset, 67
  DC, 93
Ohmmeter, 48, 62, 71
  high-power, 81
  low-power, 61-62, 81
  testing, 71
  transistor test, 48, 50
Open-circuit voltage, 60
Operational amplifiers, 46-47
  characteristics, 46
  feedback, 46
    resistor, 46
    signal, 46
Optoelectronic devices, 203-4
Oscillator, 47
  block diagram, 163
  capacitive feedback, 164
  LC circuit feedback, 164

phase shift, 164-65
  RC circuit feedback, 164
  troubleshooting, 163-64
Oscilloscope
  alternate mode, 99-100
  calibration (V) (f), 98-99
  chop mode, 99-100, 113
  component test, 103
  current measurement, 101
  digital logic, 93
  dual trace, 93-94
  frequency domain, 97
    display, 125
    r-f alignment, 157
  glitch display, 207
  graticule, 100, 126
  intensity modulation, 127
  lissajous patterns, 105
  memory, 93
  triggered sweep, 94
  use of, 10, 90-91
  vector scopes, 94

P

Phase, 55, 57
  180 degrees, 57
Phase-locked loop, 166
  motor speed control, 166-67
Phase measurement, 106
Phosphor coating, 98
Polarities, 33
Power amplifiers, 6
Power supplies, 3, 137
  battery pack, 158-59
  closed-loop regulator, 159
  connections, 30
  current regulator, 162-63
  5V, 158
  output voltage, 7
  ripple, 6, 161
  scan-derived, 162
  start-up circuit, 162
  switching regulator, 161

Power supplies *(cont.)*
  three-legged regulator, 160
  voltage, 3
Prescaler, 77
Probes, 124
  demodulator, 69–70
  detector, 242
  direct, 69
  high-voltage, 69–70
  logic, 207, 238
  low-capacity, 99
  meter, 69
Propagation delay, 195
Pulse Code Modulation (PCM), 93
Pulse Width Modulation (PWM), 161,
      170

Q

Qualitative test, 7–8
Quantitative test, 7–9

R

Receiver
  AM, 41
  antenna, 41
  converter, 42
  dead, 45
  detector, 41–42
  mixer, 42
  oscillator, 42
  r-f amplifier, 42
  speaker, 42
  tuner, 41
  volume control, 42
Recurrent sweep, 91
Resistance
  load, 15–16
Resistor, 43
  decade box, 241
  emitter, 34
  filter, 14

  model for devices, 27
  precision, 67
  substitution box, 241
  surface mount, 222
  surge limiting, 12, 14
  temperature stabilizing, 33
Rise time, 9, 96
  equation, 14
  measurement of, 14
RMS value, 67

S

Safety (cadmium), 222
Saturation, 55
Scale
  mirrored, 61
Screwdriver injection, 119
Servicing
  digital logic, 194
  microprocessor, 194
Shield, 100
Shotgun method, 136, 141, 143
Signal
  audio frequency, 45
  modulation, 45
  r-f, 45
Signal injection, 11, 117–19, 121, 132,
      243
  troubleshooting by, 121
Signal tracing, 11, 117–18, 122
Solder
  blowing on, 224
  braided wire, 227
  cold solder joint, 226
  eutectic, 221
  flux, 222
  melting temperature, 221
  sucker, 227
  types, 221
  wick, 227
Soldering, 221
  flux removal, 226
  gun, 228

iron, 227
mechanical connections, 224
tinning, 228
Spectrum analyzer, 108
Square wave
  test, 8, 14
Statistical analysis, 2, 141
Stiction, 68
Supercoolants, 186
Surface mount, 222
Sweep
  analysis, 11
  generator, 45, 124
  markers, 127
Switch, 3
  open, 4
  test, 3
Symptoms, 2, 137
Sync control, 92
Synthesizing, 168
System
  analog, 179
  digital, 179

T

Test equipment
  logic circuit, 205
  logic probe, 205–6
  logic pulser, 207
  microprocessor, 205
  troubleshooting, 3
Tester
  in situ, 164
Test jig, 50
Test signal (amplitude), 129
Tetrodes. *See* Tubes
Thermal runaway, 34
Three-state
  buffer, 200
  devices, 200
Three-terminal device, 20, 28
Time constant, 73
Time domain, 97

Timer (555), 244
Tough dogs, 118, 136
Trade-off
  gain versus bandwidth, 34
Transducer, 38, 42
Transformers, 45
  filament, 66
Transistors, 45
  bipolar, 22
  collector voltage, 82
  emitter-base short test, 84
  field effect, 72
  JFET, 22
  MOSFET, 22
  NPN, 245
  PNP, 245
  replacing, 47
  short-circuit test, 77–78
  upside-down, 36–37
Troubleshooting
  by measurement, 7–11
  Phase-Locked Loop (PLL), 168
  procedure, 19
  sequence, 122
  by shotgun method, 10
  start, 131
  statistical method, 160
Truth tables, 200, 208
Tubes
  cathode, 26
  control grid, 24, 26
  ohmmeter test, 71
  pentodes, 26
  screen grid, 24
  suppressor grid, 24
  tetrodes, 26
  triode, 24
  vacuum, 71

V

Variac, 98
VFET, 40–41
Visual inspection, 2

Voltage
  average, 101–2
  calibration, 236
  divider, 238
  drop (across diode), 15
  estimations, 83
  output, 16
  peak, 14
  polarities, 24
  proportional method, 237
  RMS, 14, 101–2, 113
Voltmeter
  accuracy, 66
  analog, 15
  test, 6
  use of, 10
  vacuum tube, 63

Volt-ohm-milliammeter
  testing capacitors with, 72
  use of, 13, 237, 240
Volume control, 42
VOM. *See* Volt-ohm-milliammeter

W

Waveforms
  square wave test, 8

Z

Z axis, 127, 131